LOOKING BEYOND THE FRONTIERS OF SCIENCE

Dedicated to the 80th Birthday of K. K. Phua

LOOKING BEYOND THE FRONTIERS OF SCIENCE

Dedicated to the 80th Birthday of K. K. Phua

Edited by

L Brink
Chalmers Institute of Technology, Sweden

N-P Chang
City University of New York, USA

D H Feng
Drexel University, USA

K Fujikawa
Riken, Japan

M-L Ge
Chern Institute of Mathematics, China

L C Kwek
Centre for Quantum Technologies, Singapore

C H Oh
National University of Singapore, Singapore

S R Wadia
ICTS-TIFR, India

 **World Scientific**

NEW JERSEY · LONDON · SINGAPORE · BEIJING · SHANGHAI · HONG KONG · TAIPEI · CHENNAI · TOKYO

Published by

World Scientific Publishing Co. Pte. Ltd.

5 Toh Tuck Link, Singapore 596224

USA office: 27 Warren Street, Suite 401-402, Hackensack, NJ 07601

UK office: 57 Shelton Street, Covent Garden, London WC2H 9HE

Library of Congress Control Number: 2022031931

British Library Cataloguing-in-Publication Data
A catalogue record for this book is available from the British Library.

LOOKING BEYOND THE FRONTIERS OF SCIENCE
Dedicated to the 80th Birthday of K.K. Phua

ISBN 978-981-126-368-2 (hardcover)
ISBN 978-981-126-369-9 (ebook for institutions)
ISBN 978-981-126-370-5 (ebook for individuals)

For any available supplementary material, please visit
https://www.worldscientific.com/worldscibooks/10.1142/13065#t=suppl

Printed in Singapore

Preface

生
日
快
乐

Looking Beyond the Frontiers of Science is the dream of all researchers, but attaining that dream is not always easy. Fermi's seminal paper on four fermion interaction was rejected by *Nature* for 'speculations too remote from reality to be of interest to the reader'. Rather than being discouraged, he submitted it to *Nuovo Cimento* and it became the cornerstone of weak-interaction theory.

The frontiers of science today continue to be advanced by the dogged determination of many researchers who think 'outside-the-box'. They encounter similar obstacles that can hinder the progress of innovation.

This volume, "Looking Beyond the Frontiers of Science" is dedicated to Professor Kok Khoo Phua (known to all as KK) who opened the door to facilitate innovations and discoveries by scientists around the world.

As a young graduate of Birmingham University, under Tony Skyrme, he returned to Singapore in 1971 and established a particle physics research group at Nanyang University, and later at National University of Singapore, and published over 80 articles in his scientific career.

As a visionary scientist, his dream was to promote frontier research in Asia and make Singapore into a hub for international scientists to gather and exchange news and ideas. In 1978 he organized an International Meeting on Frontiers of Physics in Singapore with 180 participants from around the world. He went on to organize many, many international conferences and workshops not just in Singapore, but elsewhere in Asia-Pacific.

His dream, however, was larger than simply organizing conferences and workshops. His desire was to make Singapore into a world center of publication of books and journals in frontiers of science. For this, he sought the advice of Professors C. N. Yang, Abdus Salam, and S. Weinberg. And in 1981, he took the bold step to establish the World Scientific Publishing Company. As they say in folk lore lingo, the rest is history.

This unique volume on the occasion of his 80th birthday consists of two parts.

Part I is a compilation of tributes from his prominent physicist friends around the world who have known him for decades.

Part II is a collection of research papers that celebrate his visionary approach to promote science research and education worldwide. The papers by young and old colleagues working on extending the Frontiers of Science include in this volume: Beyond the Standard Model, Dark Matter, Phase Transitions, Quantum Computing, Quantum Gravity, and String Theory.

When he was elected a Fellow of the American Physical Society (APS) in 2009, the citation read: 'For tireless efforts to strengthen scientific research throughout Asia and promote international physics education and scholarly exchanges, and for enriching science and education through the World Scientific Publishing Company he founded'.

May his efforts continue as he crosses yet another milestone!

Editors
July 2022

Contents

生
日
快
乐

Part II — Research Papers

Part I

Tributes

My Association with WSPC and Collaboration with K.K. Phua

生
日
快
乐

Ignatios Antoniadis

Laboratoire de Physique Théorique et Hautes Energies,
Sorbonne Université, CNRS, 4 Place Jussieu, 75005 Paris, France
antoniad@lpthe.jussieu.fr

In this short note, I describe my association with World Scientific Publishing Company (WSPC) and its chairman and Editor-in-Chief K.K. Phua on the occasion of his 80th birthday. My collaboration with KK was also in the context of his past capacity as director of the Institute of Advanced Studies (IAS) of Nanyang Technological University (NTU) of Singapore, trying to promote High Energy Physics in the general Asia-Pacific region and build solid connections with renown institutions in Europe, USA and the rest of the World.

My first direct interaction with K.K. Phua and his research group was about twelve years ago, in February 2010, during my trip to Singapore as an invited speaker at the International Conference in Honour of Murray Gell-Mann's 80th birthday, co-organized by Harald Fritzsch at the conference center of IAS in the NTU campus. Coming to Singapore for the first time, I was greatly impressed by the organization arrangements, facilities and general local infrastructure, which we do not see often in institutions with similar activities in the West, besides the scientific excellence of the conference from all aspects.

Since then, I became a regular guest and visitor of IAS-NTU and was participating in its conferences almost every year before the COVID pandemic period. In particular, I was a speaker for several editions of the CERN-IAS school on High Energy Physics and had fruitful interactions with KK and other members of his group. As the global situation is now improving, I look forward to visiting Singapore again and WSPC offices and meeting colleagues, collaborators and in particular K.K. Phua.

While my association with KK has been only twelve years, World Scientific journals *International Journal of Modern Physics A* (IJMPA) and *Modern Physics Letters A* (MPLA) were known to me much before. Indeed, my first publication in *MPLA* was thirty years ago, in 1992 [1]. Presently, I already have twenty publications: seven original research articles [1–7], one review based on lecture notes [8] and having received over a hundred citations in SLAC database, and twelve conference proceedings [9–20].

Furthermore, since 2010, I have served as Managing Editor of the Editorial Boards of *IJMPA* and *MPLA* which has allowed me to know better their organization, working mode and editorial challenges. I could also appreciate the considerable growth over the years with continuous improvements and successful evolution despite the worldwide competition with other scientific journals in the same research areas. Obviously, the editorial work has strengthened my association a lot with WSPC. One of its assets that I find very important is the large variety of scientific books of extremely high quality, and for different levels of readers that are being published at a constant rate. My personal library is already full of World Scientific books which I use often as reference for teaching and for my research.

The creation of WSPC about 40 years ago was part of KK's great vision to promote and develop fundamental science research in Asia and in particular in the southeast Pacific region. Indeed, the rapid and spectacular growth of these geographic areas in all aspects of life and society (financial, technological, industrial, sociological) asks for an upgrade of the higher educational system and building up fundamental research infrastructures. In this context, a local (Asian) publishing company, such as WSPC in Singapore, plays a very important role which is expected to grow further in the future.

In conclusion, KK's contribution in promoting science and fundamental research, such as High Energy Physics, in the general Asia-Pacific region and in establishing solid collaborations and strong connections with the scientific community all over the world, is remarkable.

My best wishes to KK for his 80th birthday!

References

[1] I. Antoniadis and S. D. Odintsov, Conformal dynamics of quantum gravity with torsion, *Mod. Phys. Lett. A* **8** (1993) 979–986.

[2] I. Antoniadis and B. Pioline, Higgs branch, hyperKahler quotient and duality in SUSY N=2 Yang-Mills theories, *Int. J. Mod. Phys. A* **12** (1997) 4907–4932.

[3] I. Antoniadis and G. Savvidy, Conformal invariance of tensor boson tree amplitudes, *Mod. Phys. Lett. A* **27** (2012) 1250103.

[4] I. Antoniadis and G. Savvidy, Extension of Chern-Simons forms and new gauge anomalies, *Int. J. Mod. Phys. A* **29** (2014) 1450027.

[5] I. Antoniadis and S. Cotsakis, Topology of the ambient boundary and the convergence of causal curves, *Mod. Phys. Lett. A* **30** (2015) 1550161.

[6] I. Antoniadis, S. Cotsakis and K. Papadopoulos, The causal order on the ambient boundary, *Mod. Phys. Lett. A* **31** (2016) 1650122.

[7] I. Antoniadis, Y. Chen and G. K. Leontaris, Inflation from the internal volume in type IIB/F-theory compactification, *Int. J. Mod. Phys. A* **34** (2019) 1950042.

[8] I. Antoniadis and K. Benakli, Large dimensions and string physics in future colliders, *Int. J. Mod. Phys. A* **15** (2000) 4237–4286.

[9] I. Antoniadis, D-brane physics at low energies, *Int. J. Mod. Phys. A* **16** (2001) 866–879.

[10] I. Antoniadis, Experimental signatures of strings and branes, *Int. J. Mod. Phys. A* **21** (2006) 1657–1669.

[11] I. Antoniadis, Aspects of string phenomenology, *Int. J. Mod. Phys. A* **25** (2010) 4727–4740.

[12] I. Antoniadis, From extra-dimensions: Multiple brane scenarios and their contenders, *Int. J. Mod. Phys. A* **25** (2010) 5817–5845.

[13] I. Antoniadis, Mass hierarchy and physics beyond the Standard Theory, *Int. J. Mod. Phys. A* **29** (2014) 1444001.

[14] I. Antoniadis, Quarks and a unified theory of Nature fundamental forces, *Int. J. Mod. Phys. A* **30** (2015) 1530015.

[15] I. Antoniadis and K. Benakli, Extra dimensions at LHC, *Mod. Phys. Lett. A* **30** (2015) 1502002.

[16] I. Antoniadis, Scale hierarchies and string cosmology, *Int. J. Mod. Phys. A* **32** (2017) 1730012.

[17] I. Antoniadis, Inflation from supersymmetry breaking, *Int. J. Mod. Phys. A* **33** (2018) 1844021.

[18] I. Antoniadis, Scale hierarchies, supersymmetry and cosmology, *Int. J. Mod. Phys. A* **33** (2018) 1845003.

[19] I. Antoniadis and A. Chatrabhuti, Challenges in string and supersymmetric cosmology, *Int. J. Mod. Phys. A* **36** (2021) 2044030.

[20] I. Antoniadis and K. Benakli, A de-Sitter weak gravity conjecture, *Int. J. Mod. Phys. A* **36** (2021) 2141006.

KK, Particle Physicist, World Scientific and the Institute of Advanced Studies

生
日
快
乐

Lars Brink

Department of Physics,
Chalmers University of Technology, S-412 96 Göteborg, Sweden

The Beginning

In 1978, the first Rochester conference was to be held in Tokyo. Those conferences that are biannual gather all the relevant particle physicists in the world, and are very influential for the development of the subject. There was enormous enthusiasm to go there among young particle physicists. It was expensive and there were a limited number of invitations going out. I was lucky to be one of Sweden's participants. Some months before the conference I got an invitation from K.K. Phua to pass through Singapore on my way there for a pre-meeting. I was very impressed to be invited, but had to decline since I really could not afford it. I had seen KK's name in particle physics and this was the first time I had connected him to Singapore, and perhaps the first time Singapore was being put on the physics map.

In the early 80s, we heard rumors that a new publisher was being set up in Singapore and then World Scientific was established. Early on, I was asked if I could be a main editor for Europe for the two journals *IJMPA* and *MPLA* which I gladly accepted, especially since I knew that Abdus Salam was a strong force between the idea. He was certainly one of my heroes. Since then I have been involved in the journals and the company and I noticed immediately that all the executives, assistants and personnel were highly professional and it was a delight to work with them. I am so happy to see how WSP has developed into one of the leading publishers in the world.

It's no wonder that Doreen and KK have been regarded as some of the most successful entrepreneurs in Singapore.

The next step in putting Singapore on the physics map was to organize the Rochester conference in 1990 in Singapore. This was a masterstroke by KK to manage being accepted as an organizer. The conference came a year after LEP had started, and on the theory side this was still the peak period of Superstring Theory. I was asked to organize the session on Superstring Theory and it was easy with the fantastic staff that KK provided for the meeting. Singapore was on its way up as an economic power and there was still a lot of cheap (fake) goods to buy. Many of us participants had to buy an extra suitcase to bring the purchases back. Singapore really stuck on our minds as a most interesting place in the world. When I started to come back, some fifteen or so years later, that Singapore was gone and had developed into one of the most advanced states in the world. Singapore has never stopped to surprise and attract me.

Institute of Advanced Studies

KK never lost his interest and love for physics even though the work with WSP was very demanding. Almost twenty years ago, he could launch another dream to set up Institute of Advanced Studies at Nanyang Technological University in Singapore. With a generous grant from the Lee Foundation and a strong commitment from the university, he established a new meeting forum in this part of the world. In 2018, the Institute had established itself as a major hub for high-level conferences, workshop, public talks and programs to promote science both to experts and to locals. This was a marvelous achievement. When a new president came in 2018, he changed the direction of the Institute and we all left. I had got involved in it when I had retired from my chair and from my role in the Nobel Committee, and spent many lovely months for some five years in Singapore as a visiting professor at the institute. I had free reign to invite people to the conferences and the support from KK was always 100%. The conferences were also broadcast on Youtube, so they could be followed from anywhere on the globe. These years were among the best in my life and KK and the people at the Institute and the people of Singapore all contributed to it. I still get many questions from people around the world if there will be more such conferences like the ones we organized in those years.

KK

I know how quickly a life goes on and when you reach the end of it, you often reflect on what you did and what should have been done differently. When KK now turns 80, I think he could reflect on the right things he did all through his path of life. He could have tried to stay in the limelight of particle physics seeking a position in a high level institution in the US or Europe, but he chose to go back to his home country, to his roots, to serve it. Even back home he could and did continue to do successful research, but his legacy will be what he has done for physics and science at large through his entrepreneurial and organizational skills. These are really great contributions to science, to Singapore and to the world. I will especially mention his willingness to publish memorial volumes for outstanding physicists. These are no cash cows for WSP but they are important since memories fade away quickly, however, books remain. We are all impressed and admire him for what he has done. Our only hope now is that he can continue as he did while at the same time he can enjoy the rest of his life.

To KK, My Dear Friend and Colleague

Ngee-Pong Chang

Department of Physics,
City College, City University of New York,
New York, NY, 10031, USA
nchang@ccny.cuny.edu

1. Introduction

It is such an honor and privilege to have been invited by World Scientific Publishing Company (WSPC) to be part of a team of editors to put together a tribute to my dear friend and colleague, Professor Kok Khoo (KK) Phua on the occasion of his 80th birthday. According to the Chinese Zodiac, KK was born a horse, and he has been rearing to go ever since his youth. After his high-school education in Singapore, he went overseas for higher education at University of Birmingham. His 1970 Ph.D. thesis was under Tony Skyrme.

After a year's post-doc at Niels Bohr Institute, he returned to Singapore and joined the physics faculty at Nanyang University.

2. Meeting KK 50 years ago at Nantah

And that is when I met him during my visit back to Singapore for our Christmas vacation at the end of 1971. After welcoming the new year in 1972, I went to visit Nanyang University (Chinese name Nantah) because of my fond memory of Nantah. As a young teenager I had gone with my parents to witness the Nantah opening ceremony in March 1956 when my cousin was among the first entering class there. We were all so proud that

Nantah was built on 500 acres of land in western Jurong area, entirely on the donation by Singapore Hokkien Association and with the generous support of the ethnic Chinese community in Singapore and Malaya.

So during the first week of the new year 1972, I dropped in on their physics department, and found an enterprising KK who as a young Assistant Professor had already built an impressive particle physics group, with a High Energy preprint library in the department, and with the personal collection of *Physical Reviews* donated by Prof. C. N. Yang.

Since then, we have kept up a close relationship as we both, having been born and raised in Singapore, share the desire to make Singapore more than a *'little red dot'* in the map of Asia.

3. KK's Vision

KK had a vision for making Singapore into a hub for international scientists to gather and exchange news and ideas. And as a good horse, rather than dream about the vision and wait for it to come true, he set about making that vision a reality. In making summer trips to Stony Brook (1972), Trieste (1973), Melbourne (1974), Chicago and Argonne National Lab (1976), Yukawa Institute (1977), he built a personal and lasting relation with many leading scientists around the world.

In 1978, KK successfully organized an International Meeting on Frontiers of Physics in Singapore, with 180 participants from around the world. The meeting was held at the National University of Singapore on August 14–18, 1978. Among them was Abdus Salam who gave a plenary session on Weinberg–Salam Unified Theory. This was **before** they were awarded the Nobel Prize in 1979.

To make this meeting memorable, KK managed to convince the National Academy of Singapore to publish a 2-volume Proceedings of the meeting.

But this publication, by itself, was not a viable solution for the many other international workshops and conferences that KK had in mind. Conference proceedings should be published so that those of the scientific community who did not or could not attend may yet be kept abreast of new results and discussions at the meeting.

4. Birth of WSPC 1981

KK had the deeper vision and desire to make Singapore into one of the world centers for the publication of books and journals in frontiers of science. The birth of quantum physics was largely fostered and nurtured by the publication of key innovations and ideas in European periodicals, particularly German journals, *Annalen der Physik, Zeitschrift für Physik*. But by mid-20th century, as new centers of physics were developed in America, the periodicals of American Physical Society joined the European periodicals in facilitating publication of research.

KK's desire was to enable more researchers from the rest of the world to join in looking beyond the frontiers of science and have access to publish innovations and results. We all know the story of the famous Indian physicist, S. N. Bose, how he, as a lecturer at University of Dhaka, had difficulty publishing his innovative approach to deriving the Planck radiation formula, and it was only by his direct appeal to Einstein that his paper could be published in German (after translation by Einstein). With Einstein's comment and additional insight through a subsequent paper, this resulted in what we now call Bose–Einstein statistics.

KK consulted with C. N. Yang and Abdus Salam on his idea to make Singapore a center of advanced publications of research ideas and results from the scientific community not just in Europe and Asia, but the rest of the world too. Both Yang and Salam enthusiastically endorsed his proposal to start World Scientific Publishing Company in 1981. As they say in folklore lingo, the rest is history.

5. Conferences and Workshops in Singapore

With the launch of WSPC, and encouraged by the success of the conference in 1978, KK went on to organize with the help and advice of Akito Arima the 1st Asia-Pacific Conference in Singapore on June 12–18, 1983. This conference was co-sponsored by the Physical Society of Japan, the Institute of Physics (Singapore), the National University of Singapore, the South-East Asia Theoretical Physics Association. This time, there was well over 340 participants from Asia and Pacific countries as well as America and Europe who attended the conference. And by now, KK's stature in Singapore had risen, with recognition and support of the government. During the conference, some of us participants were invited to a state dinner held on June 13 at Temasek, Istana, the official residence and office of the President of Singapore and office for the Prime Minister and Senior Minister. The dinner was hosted by then Senior Minister of Education, Dr. Tony Tan, and the honored guests included C. N. Yang, S. Chandrassekhar, S. C. Jain, H. Feshbach, C. A. Hurst, A. Arima, T. Goto, Z. Maki, C. W. Woo, P. Y. Hwang, D. H. Feng and others.

In 1987, KK organized the 1st Asia Pacific-Workshop on High Energy Physics Superstrings, Anomalies and Field Theory held on June 21–28 at a seaside resort Hotel Meridien Changi. About 340 participants came from Asia and Pacific, as well as Europe and America. In a way, it was a precursor to the bucolic setting of Nanyang Executive Centre at Nanyang Technological University, so that participants were 'secluded' far away from the city center, with the opportunity to interact and exchange ideas and build relations.

But KK did not stop with these Asia-Pacific workshops in Singapore. In 1990, he convinced IUPAP to hold the 25th International Conference on High Energy Physics in Singapore. For this he had the kind assistance and advice from Y. Yamaguchi, who arranged for a preparation meeting at KEK Tsukuba during our USA Thanksgiving recess, November 23–26, 1989. And, on August 2–8, 1990, the 25th ICHEP Conference was held at the historic Raffles City in downtown Singapore, with Yang, Nambu, Gell-Mann, Marshak, Bjorken and many other physics dignitaries in the audience.

What made it even more memorable was a historical perspective on past Rochester Conferences with Marshak showing slides from the 7th International Conference on High Energy Physics held in 1957 in Rochester, NY.

6. OCPA Conferences: Looking to the 21st Century

There were many other conferences that KK helped organized, including the series of Overseas Chinese Physics Association (OCPA) international conferences held in China, Hong Kong, Singapore, Taiwan. KK's goal has always been to promote cooperation and collaboration between researchers in Asia and America and Europe. And with the formation of OCPA in 1990, he was active in raising funds and helping OCPA organize conferences that bridge the Pacific and Atlantic oceans.

The resulting conferences are documented in the tributes by C. S. Lam, B. L. Young[a] and by T. N. Chang[b]

7. IAS@NTU

But this personal tribute cannot end without mentioning the remarkable role that KK played in establishing the Institute of Advanced Studies (IAS) at Nanyang Technological University (NTU) in 2005.

For it is not sufficient to only be hosting these memorable conferences and workshops for a few days each year, but there needs to be an Institute of Advanced Studies (IAS) with an Asian perspective so as to stimulate and promote more interest in physics research in Singapore and surrounding nations. KK's vision was to establish an IAS with facilities to invite overseas scholars to give lectures at the university and to work with post-docs. To realize this vision, KK succeeded in securing generous long-term funding from the Lee Foundation, and support from the Ministry of Education. So in 2005 the IAS@NTU was formally established.

Fig. 1.   Education Minister, Mrs. Yang, C. N. Yang, Guaning Su (NTU President), Bertil Andersson (Provost), KK.

---

[a]See tribute by N. P. Chang, C. S. Lam, B. L. Young, "KK role in launching OCPA Conferences: Looking to the 21st century".
[b]See tribute by T. N. Chang, "Kok Khoo Phua and OCPA".

As 2005 was the World Year of Physics, KK organized a special Symposium on Looking to the next 100 years of Physics and its impact on Engineering, Life Science and Technology, featuring the Asian Perspective. It was held on August 10–12, 2005 at Nanyang Executive Centre (NEC), NTU, with its own housing, dining and meeting rooms in a secluded part of the campus. The plenary speakers who covered the whole range of Engineering, Life Science and Technology were Robert Laughlin (1998 Nobel Laureate), Alexander Pines, Sankar Das Sarma, Zhi-Xun Shen, Akira Tonomura, Maw-Kuen Wu.

In 2007, IAS@NTU also organized a Conference in Honor of C. N. Yang's 85th Birthday: Statistical Physics, High Energy, Condensed Matter and Mathematical Physics, October 31–November 3, 2007).

Over the years, IAS@NTU has hosted Lars Brink, Harald Fritzsch, Kerson Huang, and others. Kerson in particular visited IAS@NTU for durations of 3 to 6 months over a period of 7 years, and started a research group with Dr. Hwee-Boon Low, Dr. Chi Xiong and Dr. Michael Good. He also taught courses which greatly benefited graduate students and researchers.

Harald Fritzsch organized yearly Conferences and Workshops on Flavor Physics in the LHC (Large Hadron Collider) Era. At the first such conferences held on November 8–12, 2010, a representative of the Directorate Office CERN, Emmanuel Tsesmelis, and the Deputy Spokesperson of CMS (Compact Muon Solenoid) experiment, Albert De Roeck, showed up. The sessions laid the seed for subsequent yearly workshops.

Beginning in 2012, IAS@NTU with the help of Fritzsch, Tsesmelis and De Roeck, organized a series of annual IAS–CERN schools and workshops that enabled and encouraged many Singapore students to go to CERN for their summer school or work at CERN by joining in the CMS experiment at LHC.[c]

Besides these Flavor Physics in LHC era, KK also organized special commemorative Conferences[d] to celebrate 60 years of Yang–Mills Field Theory, 75 years since the historic Solvay Conference, 90 years of Quantum Mechan-

[c]See tribute by Albert De Roeck and Emmanuel Tsesmelis, "Singapore and CERN".
[d]For the full list of IAS conferences visit *https://www.ntu.edu.sg/IAS/Publications* and look for 10-Year Report under the tab labelled IASReports.

ics, and to celebrate the 80th birthday of Murray Gell-Mann, 90th birthday of Freeman Dyson, 90th birthday of Rudolph Marcus, and a special memorial for Salam's 90th birthday. A general feature of these Conferences was KK's invitation to the high school students to public lectures by Nobel Laureates David Gross, Gerard 't Hooft, Robert Laughlin, Anthony Leggett, Carlo Rubbia, Cohen Tannoudji, Sam Ting, Frank Wilczek.

My own memory of these conferences is of my friend, Gerard 't Hooft, who took special care to attend the parallel sessions of reports by other participants in the field of Quantum Field Theory, and that his questions and comments helped to encourage their research. This was beyond being accosted at the many coffee breaks during the plenary and parallel sessions. NEC being in a secluded corner of the vast NTU campus, with its own housing, dining and meeting rooms was an ideal setting for much more discussions and dialogs than at a formal international conference held overseas. Participants get to have meals and coffee and tea breaks together, and the Q&A sessions after each talk were very helpful to one and all. Even as you sit in the audience and hear the announcement of new results from LHC, you might wonder what if Higgs is part of a larger family of bosons? Or on hearing many inspiring lectures on the Coming Revolutions in Theoretical Physics, you wonder what is the fundamental nature of the Vacuum?

8. Happy Birthday, KK

To conclude, I want to thank KK for the many years of friendship and collegial meetings in helping to realize his revolutionary vision. Along the way, we have had so many meals together, and each time we get to sip the special Teochew Pu'er Tea that KK brings to the restaurant.

KK, your vision has brought brightness and hope to so many younger generations of scientists around the world.

We all join in sending our best wishes to you, and may some of that brightness shine right back at you and give you joy and satisfaction in the many years ahead.

KK's Role in Launching OCPA Conferences: Looking to the 21st Century

生
日
快
乐

Ngee-Pong Chang

City University of New York, USA
nchang@ccny.cuny.edu

Chi-Sing Lam

McGill University, Canada
lam@physics.mcgill.ca

Bing-Lin Young

Iowa State University, USA
young@iastate.edu

It is a great honor for us, as founders of Overseas Chinese Physics Association (OCPA, now known as International Organization of Chinese Physicists and Astronomers) to pay special tribute to KK for his help in initiating a series of Joint Meeting of Chinese Physicists and Astronomers every 2 to 3 years.

Since OCPA was founded in 1991, we worked exclusively with the American Physical Society (APS) and organized Physics Without Border sessions at the annual APS April meetings in Washington, DC. The sessions were held under the auspices of the APS Forum on International Physics and highlighted noteworthy research by younger generation of researchers from US and overseas. In 1992, we established the Outstanding Young Researchers Award (OYRA) and with the gracious cooperation of Professors Lee and Yang, Paul Chu and Patrick Lee selected Shou-Cheng Zhang as the first

OYRA recipient. At that time, S. C. Zhang was indeed a young researcher at IBM Almaden Center after his PhD from Stony Brook. He went on to make phenomenal breakthroughs in theoretical condensed matter physics that led to prediction of topological insulators,[1] the first realistic quantum spin Hall material in spintronics, and high-temperature superconductivity. We truly believe that he was in line for a Nobel Prize in Physics, were it not for the tragedy that ended his life in 2018.

C. N. Yang,   Shou-Cheng Zhang,   T. D. Lee,   Paul Chu,   Patrick Lee

Fig. 1.   OYRA award 1992 at APS April meeting in Washington DC.

In 1993, with funding support from KK, we instituted the Achievement in Asia Award (AAA) to recognize and encourage frontier research there. And OCPA was pleased to announce the first AAA recipient at our OCPA meeting on April 12, 1993 at the annual APS meeting in Washington, DC. At the Physics Without Border session held under the auspices of the Forum on International Physics, the award was given to the promising young researcher, Zhongcan Ouyang from Institute of Theoretical Physics, Chinese

---

[1]To add a historical footnote: S. C. Zhang was invited to the 2017 Conference on 90 years of Quantum Mechanics, organized by KK, held at the Institute of Advanced Studies (IAS) at Nanyang Technological University (NTU), Singapore. There he gave a talk on the discovery of the Chiral Majorana Fermions, and in that talk, he thanked KK for the 2006 Workshop on Spintronics held at IAS@NTU where he met and discussed with Molenkamp and developed the idea of band inversion leading to topological insulator.

Academy of Science (CAS). Zhongcan Ouyang was later elected Academician of Chinese Academy of Science in 1997, and has become a world leader in the field of biological liquid crystals.

KK felt however that there should be an international conference held in China so as to benefit many more physicists in China, HK, Taiwan and SE Asia. For this, he undertook the necessary fundraising to make this idea a reality. He approached the Li Kashing Foundation in HK, and succeeded in persuading Li Kashing to fund the first OCPA international conference at Shantou University in his home-town.

So, with the help of KK, we began to prepare for the First International Conference of Chinese Physical Societies conference held in Shantou, China August 5–9, 1995. For this, he arranged for B. L. Young and N. P. Chang to make a site inspection visit to Shantou in March 1994 to check out their conference and guest housing facilities. Upon return to the US, we set up in quick succession, the Organizing Committee, International Organizing Committee and the Program Committee. We met at Chicago's O'Hare Airport on four separate occasions, June and November 94, April and June 95. The members included Jason Ho (Ohio State U), B. L. Young (Iowa State), Ling-Fong Li (Carnegie-Mellon), Sam S. M. Wong (U Toronto), C. S. Lam (McGill), and N. P. Chang (CCNY).

When the OCPA conference finally came to fruition in Shantou, August 5–9, 1995, a total of over 400 participants came for the occasion, including roughly 250 from Mainland China, 100 from outside of Asia, 30 from Hong Kong, 20 from Taiwan, and 10 from other regions in South-East Asia.

The occasion was made even more historic by the presence of all four Nobel Laureates of Chinese origin, T. D. Lee, Y. T. Lee, Sam Ting and C. N.Yang; Presidents of Academia on both shores of the Taiwan Straits, Y. T. Lee and Guang-Zhao Zhou (K. C. Chou); Academicians; OCPA and APS prize recipients; leading researchers in many frontier areas of physics; and scores of graduate students. Ernest Henley, APS President, also came. The scientific program included 12 plenary talks, 26 parallel sessions totaling about 260 talks, and poster sessions. In addition to its scientific importance, the conference was also a reunion of ethnic Chinese physicists.

Y. T. Lee   C. N. Yang   K. C. Chou

Fig. 2.   1995 OCPA1 Plenary Session.

T. D. Lee   S. C. Ting

Fig. 3.   1995 OCPA1 Plenary Session.

In addition to research results, two panel discussions, one on Science Policy and one on Physics Education, and two special sessions, one for high school students and one on Future Activities, were held. The four Chinese medalists of the 1995 Physics Olympiad as well as more than eight hundred students from the Shantou area attended the High School Student Session. The panelists of the session was organized by Kai-Hua Zhao and Da-Hsuan Feng with panelists consisting of Paul Chu, Y. T. Lee and C. N. Yang. The Science Policy Panel was organized and moderated by Ngee Pong Chang, the Education Panel by Sam Wong, and the Future Activities Session by Chi-Sing Lam.

All of us who were at this OCPA1 conference came away convinced that it has laid the groundwork for a bridge of collaboration and cooperation across the oceans. KK's vision led to a series of international conferences that continued through the years. With his help, and the cooperation of the Academia Sinica in Taiwan, we held OCPA2 in 1997 at the Institute of Physics, Taipei, and dedicated it to the 90th birthday of Professor Ta-You Wu.

OCPA2 was held in the Academia Sinica in Taipei, August 11–15, 1997. It was co-hosted by the Institute of Physics of the Academia Sinica. As in OCPA1, the four Chinese Nobel Laureates were present, T. D. Lee, Y. T. Lee, Sam Ting, and C. N. Yang, together with the Field Medalist S. T. Yau, and Steven Chu. Although not yet a laureate at the meeting, Steven Chu would later be awarded the physics Nobel prize of that year.

C. S. Lam,   Akito Arima,   Steve Chu

Fig. 4.   At OCPA2 Conference.

Another invited plenary speaker was Akito Arima, at that time the director of RIKEN in Japan, having already served as the President of the University of Tokyo earlier. What is interesting is that both Chu and Arima would later become cabinet ministers, Chu as the secretary of energy in U.S.A., and Arima as the minister of education in Japan. A picture was taken when the two stood together at the conference.

As in OCPA1, delegates came from all over the world. The delegates from Mainland China were led by Xide Xie, president of Fudan University. Unfortunately she passed away not long after the conference. A memorial session was held in her honor in OCPA3.

C. N. Yang lecture   Y. T. Lee, T. Y. Wu, C. N. Yang

Fig. 5.   OCPA2 session in honor of Prof. T. Y. Wu.

OCPA2 was timed to coincide with Prof. Ta-You Wu's 90th birthday. Prof. Wu, sometimes known as the father of Chinese modern physics, had served as the President of the Academia Sinica where the conference was held. He was the Bachelor thesis adviser of C. N. Yang, and was responsible for bringing T. D. Lee and Guangya Zhu to study in the U.S.A. after WWII.

A special session was set up in OCPAII to celebrate Prof. Wu's 90th birthday. Prof. C. N. Yang did the presentation, as recorded in the accompanying slide.

KK with Paul Chu and Leroy Chang   KK with T. Y. Wu

Fig. 6.   KK at 1995 OCPA1 and 1997 OCPA2.

After these successful events, KK continued to be actively involved in helping OCPA and proposed that OCPA3, the third International OCPA conference, be held in Hong Kong, and offered to help with the fund-raising for the meeting and the OCPA executive council accepted his proposal.

So OCPA3 was held July 31–August 4, 2000 at the Chinese University of Hong Kong. Apart from the many plenary and parallel sessions devoted to the different areas of frontier research in physics, OCPA3 also dedicated plenary sessions in remembrance of Profs. T. Y. Wu, C. S. Wu, and X. D. Xie, President of Fudan University.

On the opening day of OCPA3, a special Memorial Session was held in honor of Prof. Ta-You Wu, where Prof. Kun Huang, T. D. Lee, C. N. Yang, and Chun-Shan Shen gave moving tributes to Prof. Wu's influence on the many generations of physicists. At KK's suggestion, the session concluded with a Q&A session that invited the many young students sitting at the upper floor to pose questions to T. D. Lee and C. N. Yang on how to pursue a meaningful career in physics. Their questions were in the context of cross-cultural backgrounds that could be compared with the young researchers in America and Europe.[2]

Kun Huang, T. D. Lee, C. N. Yang, Chun-Shan Shen

Fig. 7.   2000 OCPA3: TY Wu Memorial Session group photo.

This outreach to the younger generation of students continued to be a feature of the public lectures and sessions at the many international conferences not just at OCPA but also subsequently at the workshops and confer-

[2]See "Commemorating the Past and Looking towards the Future", p. 53–54, World Scientific Publishing Co., 2002.

ences held at the Institute of Advanced Studies at Nanyang Technological University (IAS@NTU), of which KK was the founding director.

For the record, we list the continuing series of International OCPA Conferences in the historical order:

- OCPA1   Aug 5–9, 1995   Shantou
- OCPA2   Aug 12–15, 1997   Taipei
- OCPA3   Jul 31–Aug 4, 2000   Hong Kong
- OCPA4   Jun 28–Jul 1, 2004   Shanghai
- OCPA5   Jun 27–30, 2006   Taipei
- OCPA6   Aug 3–7, 2009   Lanzhou
- OCPA7   Aug 1–5, 2011   Kao-Hsiung
- OCPA8   Jun 23–27, 2014   Singapore
- OCPA9   Jul 17–20, 2017   Beijing

We at OCPA have truly been blessed by the vision of KK in launching and helping to support these continuing and memorable international conferences.

As part of this tribute to KK, we conclude with some photos of KK at our OCPA meeting at the 1997 annual April meeting of APS.

KK,   C. S. Lam,   T. K. Lee,   J. C. Peng
N. P. Chang,   L. F. Li,   B.L. Young

Fig. 8.   1997 April 18 OCPA meeting in Washington, DC.

KK with

B. L. Young, L. N. Chang, C. Y. Wong, N. P. Chang

Doreen with on the right

B. L.'s wife and L. N.'s wife

Fig. 9. 1997 April 18 OCPA meeting at Washington DC.

To KK and Doreen, enjoy this walk down memory lane!

To the rest of us, we conclude this tribute with the quotation 'Yinshui Siyuan' that applies to each and every one of us, to be mindful and ever grateful for the source of inspiration and knowledge from our teachers as we look to Beyond the Frontiers of Science.

Yinshui Siyuan

When you drink living water,

Be grateful for the source

Fig. 10. T. D. Lee calligraphy.

Kok Khoo Phua (潘國駒) and OCPA

生日
快乐

Tu-Nan Chang (張圖南)

Department of Physics and Astronomy,
University of Southern California, Los Angeles, CA 90089-0484, USA

It gives me great pleasure to have this opportunity to congratulate Kok Khoo and wish him the very best on his 80th birthday, and for many more years to come. Kok Khoo was involved in the founding of OCPA (*O*rganization of the *O*versea *C*hinese *P*hysicists and later renamed as the International *O*rganization of *C*hinese *P*hysicists and *A*stronomers) as a member of the original planning team, in charge of its Editorial Board for publications, working with Ngee-Pong Chang, Chi-Sing Lam, Bing-Lin Young, and others in 1990. Kok Khoo and his World Scientific Publishing Company generously donated their services, printing, and distribution of the OCPA Newsletters starting from October 15, 1990 (from Vol. 1, No. 2), reaching out to the members when OCPA was first organized.

Kok Khoo was the one who made the initial contact between Shantou (汕頭) University and OCPA and co-chaired the local organizing committee for its first international conference (OCPA1) from August 5–9, 1995 in Shantou. The conference was attended by four Nobel Laureates (楊振寧 C. N. Yang, 李政道 T. D. Lee, 李遠哲 Y. T. Lee, and 丁肇中 S. C. C. Ting) and about 400 participants from the US, Europe, China, Taiwan, Hong Kong, and Singapore, and the conference proceedings was published by World Scientific. The highlights of Kok Khoo's invaluable contribution to OCPA1 are detailed by Ngee-Pong Chang, Chi-Sing Lam, and Bing-Lin Young in their tribute to him in this volume.

C. N. Yang and Y. T. Lee at OCPA1   KK next to Paul Chu and Leroy Chang

Over the past three decades, Kok Khoo has served continuously as a member of the OCPA Executive Council and also as member of its International Organizing Committee or the International Advisory Committee for all OCPA international conferences (OCPA1–9). Kok Khoo also established the OCPA AAA (Achievement in Asia Award) Tan Kah Kee Prize through donation by the Tan Kah Kee International Society until 1999. It was Kok Khoo's dedication and enthusiasm that inspired my commitment to sponsor the OCPA AAA Robert T. Poe Prize honoring my late graduate advisor. In addition, Kok Khoo co-chaired with Nai-Chang Yeh the Organizing Committee of 2014 OCPA8 in Singapore. His participation and contribution to the OCPA8 from its initial planning to the meeting itself are detailed by Nai-Chang in her tribute to him also in this volume.

The conference proceedings published by World Scientific for the first three OCPA International meetings have recorded vividly a few of the most memorable occasions during those conferences. Some of the OCPA members may still remember the three special sessions in memory of Ta-You Wu (吳大猷), Chien-Shiung Wu (吳健雄), and Xi-De Xie (謝希德) highlighted by the Conference Proceedings for the 2000 OCPA3 in Hong Kong.

During the memorial session for Ta-You Wu, to make sure that C. N. Yang and T. D. Lee would be sitting next to each other, the members of the OCPA Executive Council and a few additional OCPA members were mobilized to occupy all other seats except the two center ones of the front row before the arrival of Yang and Lee. It resulted in the photograph shown in the OCPA3 Conference Proceeding together with an earlier photograph of Yang and Lee with Wu taken at the 1997 OCPA2 in Taipei celebrating the 90th birthday of Ta-You Wu (shown below).

In addition to Kok Khoo's continuous dedication to the OCPA, through his outreach and generous invitations, many scientific books and monographs authored by scientists with ethnic Chinese backgrounds were published by World Scientific over the years. Without Kok Khoo's encouragement, many of those authors would not have even considered publishing. These continuous efforts by Kok Khoo and World Scientific over the years have broadened the inclusiveness, impact, and contributions by Chinese scientists in various scientific fields, which were traditionally Euro-centric.

Finally, on behalf of our many fellow OCPA members, I would like to express our best wishes and gratitude to Kok Khoo for his friendship and support of OCPA on his 80th birthday.

Singapore and CERN

生
日
快
乐

Albert De Roeck and Emmanuel Tsesmelis

CERN, The European Organization for Nuclear Research,
1211 Meyrin, Switzerland

Prof. Kok Khoo Phua has always had a strong interest in CERN activities and in connecting Singapore closer to CERN. In this contribution we recall some of the history of the scientific exchange between CERN and Singapore since 2010, and in particular Prof. Phua's role in it.

1. Introduction

The authors, Albert De Roeck and Emmanuel Tsesmelis, senior scientists at the CERN laboratory, visited several countries in South-East Asia at the start of 2010, which happened to coincide with the start-up of the Large Hadron Collider, the LHC. The LHC is the largest accelerator ever built by mankind so far and its start-up was awaited with large anticipation by the scientific community.

The CERN scientists passed by Singapore, on invitation by Prof. K.K. Phua and Prof. L. C. Kwek from the the Institute of Advanced Studies (IAS) located at the Nanyang Technological University (NTU) of Singapore. This was our first direct contact with the Singapore HEP community, also involving scientists from the close-by National University of Singapore (NUS) that offers lectures and a Masters Programme for particle physics. Emmanuel was at the time a representative of the CERN Directorate Office, and Albert the Deputy Spokesperson from the CMS experiment. They were on a mission to explore opportunities of interest of countries in South-East Asia (and the Middle East and North Africa region), for collaboration with CERN.

The meeting was held at NTU in the executive and guest wing, and little did we know that this was going to be a regular point of return for the both of us during the decade that had just started.

The scientists in Singapore and Prof. K.K. Phua, in particular, have been very interested to get in close collaboration, if possible join, the activities at CERN. This has led during the last decade to a plethora of activities such as seminars, workshops, CERN related schools, as well as courses given on particle physics and accelerator technologies. The latter has led to several Singapore PhD students that have worked at CERN with CMS, and numerous students that have participated in the popular CERN Summer Student programme.

In this contribution, we will go back through the history of the collaboration with Singapore and highlight the strong impact Prof. K.K. Phua has had so far on these activities and developments.

2. CERN and Experiments

CERN is the European Laboratory for Particle Physics, the largest laboratory for particle physics in the world and home of the Large Hadron Collider (LHC), a 27 km long atom smasher located 100 m underground in the Geneva region. A few months before the first visit of Emmanuel and Albert to Singapore in 2010, the LHC delivered its first collisions, at an initial energy of around 1 to 2 TeV of centre-of-mass (CM) energy.

CERN is an intergovernmental organization created in Europe in 1954, not long after the Second World War, in order to create a new start of collaboration on the European continent, and it became very fast a place of scientific excellence. Currently, CERN has 23 Member States contributing to the running of the laboratory, and 10 Associate Member States. In total, it has about 3,500 staff members and fellows, and in particular, it receives about 15,000 scientists per year that come from all over the world to the laboratory for longer or shorter periods to take part in the CERN science programmes. Over the years, CERN has constructed several large particle accelerators for experiments to conduct fundamental scientific research. Such an accelerator and experimental programme requires cutting edge technologies and

CERN thus became a centre of excellence, for example, for vacuum technology, cryogenics, detector technologies, progress in (grid-)computing and other needed developments. CERN is also well known as the place where the World Wide Web was born in 1990, originally intended by Sir Tim Berners-Lee as a tool to facilitate the communication and document/data exchange among CERN collaborators for scientific purposes, but in fact, it revolutionized the development and use of the internet communication for the whole world community, especially since the tool was offered to be used at no charge.

CERN nowadays is a beacon for pushing back the frontiers of knowledge, developing new technologies for accelerators and detectors, a place to train the scientists and engineers of tomorrow, and a forum to unite people from different countries and cultures. A leitmotiv of CERN has always been "Science for Peace" and it had a special role connecting scientists across the world during the cold war period in the 1970s and 1980s.

The science programme and the technology know-how makes CERN an interesting partner for a country like Singapore, which K.K. Phua fully realized and he and his team were very welcoming to us when visiting Singapore in 2010. The visit triggered the start of several initiatives in Singapore to connect with CERN and set up common projects, as there were the Joint CERN-Singapore schools, providing particle physics lectures, assisting with particle physics seminars and conferences, and with the local proposal for Singapore to join an LHC experiment at CERN.

The experiment of choice to join for Singapore was the CMS experiment at the LHC (not completely de-correlated of course with the fact that Albert was a member of this first "scouting team" to come ashore in Singapore). CMS, or the "Compact Muon Solenoid" is one of the two general purpose experiments at the LHC — the other being ATLAS — which has a broad physics programme, which includes two main physics targets: "the hunt for the Higgs Boson", and "exploring the TeV energy scale for physics beyond the Standard Model". The experiment is shown in Fig. 1.

Prof. K.K. Phua initiated the idea of Singapore engaging in CMS, and was the main driver to submit a project to the Singapore funding agencies

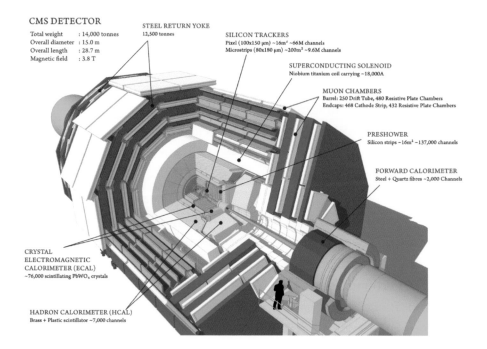

Fig. 1.   The CMS experiment at the LHC.

in order to secure funding for becoming a full member of the collaboration. Since this was likely a process that could take up to several years, it was also considered opportune to create a channel for students to join the experiment immediately to prepare a Masters or PhD thesis working with CMS physicists and with the CMS data. To that end NUS and CMS signed an expression of interest agreement (EOI) in 2013, allowing students to engage in experimental LHC physics.

Alas, after several trials, and despite the strong personal engagement of Prof. K.K. Phua with the funding agencies, so far no agreement on funding has been reached that would allow Singapore to become a full member of CMS. But efforts still continue.

3. The Joint Singapore-CERN Schools

Prof. K.K. Phua has been instrumental in launching in 2012 the IAS and CERN schools in particle physics and related areas. The first school was held on 9–31 January 2012 at NTU and attracted around 50 students [1].

Building upon the strong collaboration between IAS and CERN, the School included an outstanding programme composed of core lectures and workshops by Nobel Laureates and eminent visiting professors of IAS, as well as follow-up discussions and special seminars by the students. The physics themes included frontiers in particle physics and cosmology, the view of the cosmos with strings and branes, interdependence between physics and technology, etc.

The schools were targeted at advanced graduate students, young post-docs and faculty covering frontier topics in particle physics and cosmology, as well as the latest technology in accelerator physics. The lectures were intended to enable interested researchers to join in the scientific exploration in these exciting fields.

4. The Discovery of the Higgs Boson

A major milestone achieved at the LHC in 2012 was the discovery of a Higgs Boson, jointly announced by the ATLAS and CMS experiments on the 4th of July [2,3]. This discovery could only be achieved thanks to the excellent performance of the LHC in 2011 and especially in 2012, the superb function-ing of the experiments, the quality of the data collected by the experiments, and the well-organized data analyses procedures for this important physics topic, which engaged a large number of students and young postdocs, sur-rounded by senior researchers, producing an excellent mix that offered a very fertile forum to push these analyses to a high level of sensitivity. All that led to what would be called an 'early discovery' of the boson compared to what was anticipated prior to data taking at the LHC.

The Brout–Englert–Higgs (BEH) Mechanism was introduced in 1964 [4–7], at a time the Standard Model (SM) as we know it today still had to be fully recognized as a main contender for the particle physics theory. The mechanism would become a key ingredient in the Standard Model, and when in the early 1970s the theory was demonstrated to be normalizable such that the cross-section calculations of SM processes give sensible finite predictions the attention of the community turned to establishing the possi-ble experimental evidence for the field associated with the BEH mechanism, namely a fundamental scalar particle — allegedly coined "Higgs particle" by Ben Lee in 1972.

Early papers, e.g. the seminal paper of John Ellis, Mary Gaillard, and Dimitri Nanopoulos [8] in 1976, presented an in-depth analysis of the phenomenology of the production and decay of such a putative Higgs Boson; the one parameter that was and remained unknown till 2012 was the mass of the Boson. The paper also concludes in the discussion that "experiments should not be built for just searching for the Higgs" but the attitude to that clearly changed in the last 20 years and led to the advent of the LHC. One of the authors, John Ellis, would over the years become a regular visitor to Singapore and the IAS and a lecturer at the Singapore-CERN particle physics schools.

The most sensitive analyses that led to the discovery of the new particle were the $H \rightarrow \gamma\gamma, H \rightarrow ZZ$ and $H \rightarrow WW$ decay channels. On July 4th, both experiments reported an observation with a significance of 5σ [2,3]. Figure 2 shows a candidate Higgs decay to two photons. At the time, Albert

Fig. 2.   Candidate Higgs boson event from collisions between protons in the CMS detector on the LHC. From the collision at the centre, the particle decays into two photons (dashed yellow lines and green towers). (Image: CMS/CERN)

visited Singapore on his way to the conference in Melbourne Australia early July, to give a seminar and prepare the local scientists on "things to come".

The importance of the Higgs Boson cannot be over-estimated: This particle is of a nature that we have never seen before — a fundamental scalar particle — and its discovery has far reaching consequences also for ideas and developments of theories beyond those of the Standard Model.

5. A HEP Project for Singapore

Singapore has had a long and distinguished history in particle physics. A high-energy physics group started in Singapore in the 1970s, concentrating in theoretical physics. Some important high-energy physics conferences and workshops have been held in Singapore since then, notably the International Conference on High Energy Physics (Rochester Conference) in 1990. In 2008, Singapore organized the conference "Particle Physics, Astrophysics and Quantum Field Theory: 75 Years since Solvay". NUS and NTU have trained more than twenty PhD students in particle physics and several of them have made careers in particle physics. The high-energy physics group of Singapore is considered to be sizeable by South-East Asia standards and works closely with neighbouring countries.

Building on this strong foundation. Prof. K.K. Phua spearheaded efforts to strengthen research in high-energy physics in Singapore. A key element of this effort is to become closer to CERN through a formal framework agreement concluded in 2012 to enable scientists, engineers and students from IAS to further develop their professional training, in particular through participation in CERN's scientific and training programmes.

The principal initial domains in which training was offered includes experimental and theoretical particle physics and related aspects of technologies for particle accelerators, particle detectors and information technology.

Moreover, as part of the agreement, CERN offered to host participants from IAS in its Summer Student and other training programmes within agreed quotas, and the IAS may send their scientists, engineers and students to be trained at CERN.

Prof. K.K. Phua used the implementation of this agreement to pave the way for the conclusion of an International Co-operation Agreement with CERN, which would provide a general framework to enable the parties to broaden their co-operation, beyond the further development of professional training, to other activities at CERN, and to convince the national authorities of Singapore to support and strengthen the field of high-energy physics research in Singapore.

6. Singapore and CMS

In 2013, CMS and NUS co-signed a EOI allowing students to participate in the CMS experiment. From NUS, this involved Profs. Phil Chan Aik Hui and Oh Choo Hiap. This facilitated traveling from Singapore to CERN to become integrated in the experiment in a very smooth way. In total, three PhD theses were conducted on CMS data analyses in the past years. The pioneer, namely the first student from Singapore to come out to CERN for a large fraction of his time and engage on data analysis, was Dr. Wang Wei Yang. His scientific work was connected to making measurements of underlying event characteristics, and it got published as a CMS collaboration publication.

In 2015 the CERN connection was further tightened when NTU, under the initiative of Prof. K.K. Phua and with a lot of logistics help from NTU Prof. Lock Yue Chew, Emmanuel and Albert were offered a visiting professorship at NTU, in order to give a combined course (Introduction to Accelerators and to Experimental Particle Physics) in 2015. The combined course has been held every year since, each year attracting the order of 25 students (with a maximum of 50 in one year). In 2020 and 2021, the course had to be given remotely due to the pandemic but it did not undermine the enthusiasm of the engaging students. In fact, several of the students over the years have gotten infected by the "HEP" virus during those years and about six of them have followed up after the course with end of term projects carried out on the CERN physics programme. Also about 15 students from Singapore have come to Geneva to take part in the CERN Summer Student programme in the last decade.

One student in particular we like to single out is Ameir Shaa Bin Akber Ali from NTU. Some time in 2016, Ameir came up to Albert during a Workshop at NTU, where Albert gave a talk on "Exotica searches at the LHC", and fully confident, said "I want to discover Monopoles at the LHC. Could you tell me how I can go and do that? Albert, also a member of the MoEDAL experiment at the LHC had indeed some advice to give and the rest is history: Ameir became a CERN summer student, and then came for six months to CERN to work on a Master thesis for Singapore. He engaged in the MoEDAL experiment and became one of the early key members on the analysis of the data. After his Masters thesis the first idea was that he would follow up with a PhD thesis with us at MoEDAL, conducted from NTU. But NTU was not quite ready for that so Ameir left to follow his science interests and joined the MoEDAL group of University of Alberta, Canada, and is now completing his PhD. Whether his wish came true to discover Monopoles, we will have to wait until he delivers in his manuscript. . .

During the past decade Prof. K.K. Phua as initiated many workshops — typically a few per year — on fundamental physics via the IAS. The first one was in 2013 [9]. These workshops have been extremely useful and efficient, and the leading organizer was often Harald Fritzsch, one of the fathers of the theory of QCD. Albert and Emmanuel were typically involved in assisting with the LHC aspects of such a workshop programme. Unfortunately, due to a management change at the IAS these excellent workshops have been discontinued as of 2019, which has been a true loss for the enthusiastic community.

7. The ASEAN Federation of Physics Societies

The Association of South-East Asian Nations (ASEAN) was formed on August 8, 1967 and celebrated its 50th anniversary in 2017. Although the original aim for the formation of ASEAN revolved around the need for economic growth, social progress and social cultural evolution, the region has steadily become visible for its scientific activities in recent years.

On 29 August 2017, in the same month that commemorates the fiftieth year of the formation of ASEAN, senior physics representatives from

ASEAN came together at the Nanyang Executive Centre in NTU to endorse the Charter for the official establishment of a very-much desired ASEAN Federation of Physics Societies (AFPS). It was a fruitful meeting with the signing of the AFPS Charter and Constitution by the ten ASEAN Physics representatives at the roundtable. The event was supported by the Singapore's Ministry of Foreign Affairs. The establishment of AFPS is an important milestone as it serves as a platform to strengthen the capacity of physics education and research in the region; to facilitate the sharing of information and collaboration across the member states; to improve how resources could be accessed and distributed across the communities; and to provide a strong collective negotiating power in cooperation and collaboration with other regional and international physical societies.

The inaugural meeting of AFPS served as a platform for the formation of closer collaborations among ASEAN countries for scientific activities, and more specifically for stronger ties in physics and physics education in the region.

Fig. 3.   Opening Ceremony of the ASEAN meeting on 29 August 2017: (from left) Prof. Siaw Kiang Chou (Advisory Board Member from the ASEAN Committee on Science and Technology), Mr. Chee King Tan (Deputy Director of the ASEAN Socio-Cultural Community, Singapore's Ministry of Foreign Affairs), Prof. Kok Khoo Phua (Founding Director of IAS NTU) and Prof. Rajdeep Singh Rawat (President, Institute of Physics Singapore).

All countries are in agreement that concrete steps have to be taken to ensure that ASEAN, after fifty years in existence, should finally attain a level playing scientific field with other advanced regions and international organizations, with the most important step being that of capacity building and training of a new generation of physicists.

Moreover, the importance of continuing the pursuit of international collaborations and dialogues with international partners, as had been done since the early days of the establishment of ASEAN, was underlined.

The historic signing of the Charter established a Memorandum of Understanding between the physics communities of the ten ASEAN countries that would lead to closer collaborations in training programmes, periodic conferences aimed at promoting regional research excellence, and coordination of research and developmental priorities. Prof. K.K. Phua served as the Interim President for the formation of AFPS.

8. ASEAN Workshop on Frontiers of Physics 2017 in Partnership with CERN

The Frontiers of Physics Workshop was held on 30 August 2017 in conjunction with the inauguration of AFPS. The workshop was not merely about reporting on the future potential of cutting-edge big collaborative experiments in Europe and East Asia; nor was it about the knowledge transfer enabled by the technologies developed through research agenda at the cutting-edge. Rather, it was about sharing information about the resources, facilities and possibilities in physics that could be utilized by members of the ASEAN communities, in collaboration with other emerging and mature physics collaborations, to facilitate and bring the work they are already doing into the next level.

Even as material and energy science remain primary to the research agenda of ASEAN, the region is in the process of developing research agendas in the area of high energy particle physics, astrophysics and cosmology, space science, biophysics, and biomedical sciences, among others, that will, and are, using technologies developed as a consequence of the instrumentations built for doing big science.

At this time, the popular CERN Schools that had been taking place annually at various countries had also taken place in the ASEAN region in Malaysia, Thailand and Singapore, among others. To date, all of the ASEAN countries have developed relationships with CERN, from having initiated first contact to becoming full-fledged membership within some of the experimental collaborations. Both Malaysia and Thailand are members of the CMS collaboration while the Philippines and Singapore have direct association with CMS. Indonesia is a member of the ALICE collaboration while at least nine of the ASEAN countries are now sending summer students to CERN on a regular basis.

Fig. 4.　Participants at the ASEAN Workshop on Frontiers of Physics 2017 (in partnership with CERN).

9. Outlook to the Future

Thanks to the strong efforts and engagement of Prof. K.K. Phua, CERN has been introduced to the Singapore science community over the last decade. Singapore students have been coming to CERN for shorter or longer periods, to conduct scientific work. Singapore has a lot of potential to have a fruitful collaboration with CERN but for a sustained partnership it will be imperative that appropriate funding and academic channels to support this research branch can be consolidated. When that is achieved, Singapore has all the potential to be a strong partner and become a leader in the field in South-East Asia. When that happens one should always remember the major achievements and contributions of Prof. K.K. Phua who has ignited this process!

References

[1] First IAS-CERN School, Singapore. `https://old.ocpaweb.org/conference/IAS_CERN_rev-poster.pdf`.

[2] S. Chatrchyan *et al.*, Observation of a New Boson at a Mass of 125 GeV with the CMS Experiment at the LHC, *Phys. Lett. B* **716** (2012) 30–61.

[3] G. Aad *et al.*, Observation of a new particle in the search for the Standard Model Higgs boson with the ATLAS detector at the LHC, *Phys. Lett. B* **716** (2012) 1–29.

[4] F. Englert and R. Brout, Broken symmetry and the mass of gauge vector mesons, *Phys. Rev. Lett.* **13** (1964) 321–323.

[5] P. W. Higgs, Broken symmetries, massless particles and gauge fields, *Phys. Lett.* **12** (1964) 132–133.

[6] P. W. Higgs, Broken symmetries and the masses of gauge Bosons, *Phys. Rev. Lett.* **13** (1964) 508–509.

[7] G. S. Guralnik, C. R. Hagen and T. W. B. Kibble, Global conservation laws and massless particles, *Phys. Rev. Lett.* **13** (1964) 585–587.

[8] J. R. Ellis, M. K. Gaillard and D. V. Nanopoulos, A phenomenological profile of the Higgs boson, *Nucl. Phys. B* **106** (1976) 292.

[9] First IAS-CERN Workshop, Singapore. `https://indico.cern.ch/event/263720/`.

Twenty Years and Counting

生
日
快
乐

Berge Englert

Centre for Quantum Technologies,
National University of Singapore

Professor Kok Khoo Phua — KK for his friends — and I first met in August 2002 during my exploratory visit of the physics department at NUS. Artur Ekert had suggested that I come to Singapore for two years to help with the build-up of the quantum technology group. At the time I knew very little about Singapore and, therefore, it was prudent to explore the country and the university before committing myself.

In the course of this exploration, my wife and I were thoroughly exposed to the great Singaporean hospitality, with KK and others making sure that we got a serious dose of the splendid food of so many kinds. As a consequence, a two-year engagement in Singapore became a temptation I couldn't resist. And two became twenty . . .

Of course, I had known about World Scientific before I met KK, as many of us physicists have contributed articles to the numerous proceedings volumes published by WSPC, and was very pleasantly surprised to learn that this educated and friendly gentleman had founded this publishing house. KK's hard work — supported by his dear wife Doreen and later by their children — has turned it into one of the leading publishers of scientific books and journals. As the author of books and edited volumes in the WSPC catalog, I experienced the skills of the excellent staff assembled by KK and the same praise is deserved of the desk editors and production team of the WSPC journal that KK entrusted to me. These observations confirm the obvious: The creation of World Scientific is a major accomplishment, a conspicuous contribution to the culture of science, a legacy of lasting value.

On the occasion of his 80th birthday, I wish KK many more productive years of good health in Doreen's company with the privilege of witnessing their grandchildren grow and develop.

Yours truly,

K.K. Phua and I

Da Hsuan Feng

*Former M. Russell Wehr Professor of Physics, Drexel University
and former Vice President for Research of the University of Texas at Dallas*

It is incredible that someone who is so resilient and is constantly displaying infinite energy and creativity professionally like a young and vibrant middle-age individual can be celebrating his 80th birthyear! But then, ladies and gentlemen, we are not talking about anyone, we are talking about Dr. Kok Khoo Phua (who the world knows simply as KK)!

I am sure many world-renowned people who contributed to this festschrift for KK will claim to know him for a long time. Well, I think I can probably make the claim of knowing him the longest! In Singapore, at least when we were growing up in the 50s, children from the Guangzhou ethnic-group (which I belong to) tend to go to Guangzhou-clan-sponsored schools and Chaozhou ethnic-group (which KK belongs to) tend to go to Chaozhou-clan-sponsored schools.

For some unknown reason, KK's parents made the decision to send him to the primary school, known as Yang Zheng (养正) elementary school. Yang Zheng elementary school was sponsored by the Guangzhou-clan. There was no doubt at the time, Yang Zheng was one of the best if not the best elementary schools in Singapore. After all, how many elementary schools in Singapore today can claim to have an alumnus like Xian Xinghai, the composer of the world-renowned Yellow River Cantata! (https://en.wikipedia.org/wiki/Yellow_River_Cantata).

Well, both my elder brother Da Fei Feng, myself and KK were all students at Yang Zheng, and Da Fei and KK were in the same year, and I was two years behind them. This was in the mid-50s. Of course, since KK was known to be a spectacular student in Yang Zheng, and through my brother Da Fei I knew of him then (by reputation only), I remember I was truly in awe of him. After all, at that stage of our young lives, two years ahead in school, for all practical purposes, could just as well be infinitely far ahead!

After Yang Zheng, our lives diverged. For our next encounter, one needs to fast forward to the early 70s when I was a postdoctoral fellow at the Department of Theoretical Physics at the University of Manchester. On one of my visits to Singapore then, I tried to explore the possibility of getting a position at Nanyang University, the predecessor of the current Nanyang Technological University. It was at that visit to the campus that I was reconnected with KK. I was truly grateful to KK for his genuine enthusiasm towards me and we promised each other that we should remain in close contact, which we did.

As a theoretical nuclear physicist, I am truly impressed that KK was a doctoral student of one of the giants in my field, Tony Skyrme of the University of Birmingham in the UK. In fact, for about a decade in the 90s, when many nuclear physicists were gravitating their interest from the nucleus to the nucleons (protons and neutrons), the terminology "Skyrmion" was clearly a ubiquitous physics vocabulary!

Reflecting on what KK had done in his life and continues to do, one could tell that the impact of Tony Skyrme on him goes well beyond physics. Of course, Tony Skyrme must have bestowed into KK's psyche a vast knowledge of physics and how to carry out research in physics. However, he did something that was even more "out-of-the-box", and surely had impacted KK's approach in life.

Few people know that in Tony Skyrme's wikipedia description of him, there is a paragraph which states:

"In 1958/59 he had travelled with his wife as tourists in a one-year overland circumnavigation of the globe by car and land rover. They fell in love with the lush tropical gardens of Malaysia and

decided to settle there. In 1962 they left Harwell and Skyrme
took up a post in the University of Malaya in Kuala Lumpur."
(https://en.wikipedia.org/wiki/Tony_Skyrme)

Yes, my friends, Tony Skyrme and his wife did not just go from UK to Malaysia as "normal" people would do, namely to fly there. They drove there! It is this sort of intellectual mindset that Tony Skyrme had which I am sure profoundly influenced KK as his student.

I am sure it is not difficult for anyone who knows KK well to recognize that he is, first and foremost, a person who remembers his roots.

In nearly everything KK does, he remembers his Southeast Asian roots. Surely after he completed his Ph.D. from Tony Skyrme, a more natural progression of his career would be to seek a position in the Western world. Instead, he immediately returned to Singapore and taught physics at the then Nanyang University. There is no doubt that "you can take KK out of Singapore (for a while), you certainly cannot take Singapore out of KK!"

The fact that during these times, in Singapore, the scientific support for infrastructure was clearly lacking especially for research in high energy physics, KK took that not as a "show-stopper" but as a "challenge". In the early 70s when he first returned to Singapore, KK understood well that scientific collaboration internationally was one of the ways that he could build a robust research activity. With such a preamble, he began to collaborate with western scientists.

Furthermore, by the late 70s, he had initiated several international physics conferences in Singapore, something which was obviously a bold step in that region of the world. I remember I was invited by KK to attend one such conference. These conferences became the platform for younger generations of Southeast Asians to meet and have dialogues with some of the most notable scientists from all over the world.

For example, one such scientist was Marcel Bardon, who was a former student of the Nobel laureate Leon Lederman and was then head of the Physics Division of the US National Science Foundation. I remember distinctly how the younger people from Southeast Asia were able to meet such "world-class" scientists and to see that they were not in the least intimidating! This is what KK often told me later-on, he wanted "to lift the 3rd world mindset of the younger generations of Southeast Asians to become 1st world!"

In this vein, I would be grossly remiss if I were to talk about KK's achievements and not talk about his bold dream of building a scientifically progressive Institute of Advanced Studies (IAS) at Nanyang Technological University. In the early part of the 21st century, KK had another dream of wanting to bring the most advanced and up-to-date scientific thinking to Singapore. From his vast scientific connections, he befriended Professor Lars Brink of Chalmers Institute of Technology in Sweden to assist him to carry out this task.

In a decade and a half, KK and Brink were able to organize conferences that delved into the deep foundations of humanity's scientific thinking. One of them which I had the great honor of participating is a conference to discuss how quantum mechanics transformed humanity since it had been formally recognized as a scientific discipline in 1926.

At that conference, I saw not only top-notched scientists from all corners of the world talking about the deep implications of quantum mechanics, philosophically, technologically and of course scientifically, but also a vast number of young Southeast Asian scientists were able to rub shoulders with the speakers in a free-for-all manner. If such a cultivation of scientific thinking for humanity can be deepened in Singapore, in fact, make it a new normalcy, there is no doubt that Singapore could be a "scientific shinning city on the hill!"

As is obvious from the above discussions, whatever KK pursued, he would always have a grand intellectual scaffolding in his mind first. The rest is merely putting the right ornaments on such a scaffolding. This is KK's *modus operandi*, and I know it will continue!

KK and a Nice Memory of the Emergent Singapore

Kazuo Fujikawa

Interdisciplinary Theoretical and Mathematical Sciences Program (iTHEMS), RIKEN, Wako 351-0198, Japan

Some time in 1982 or so (my memory may not be accurate), Doreen and Kok Khoo Phua visited Prof. Saburo Miyake, then Director of the Institute for Cosmic Ray Research which was located at Tanashi in the suburb of Tokyo. Apparently, KK was an old friend of Prof. Miyake, so entertained Doreen and KK at a restaurant near the Tanashi station; somehow, Prof. Miyake invited me as well to the restaurant. This was the first time I was meeting Doreen and KK. During the conversations at the dinner, I learned that Doreen and KK had a plan to initiate a publishing house in Singapore.

After I moved to Hiroshima University in 1983, I was asked to serve as one of the editors of *Modern Physics Letters A* published by World Scientific Publishing Co. Also, some time later in the 1980s, KK organized a workshop in Singapore for which I was invited together with some other Japanese physicists. My impression of Singapore which I was visiting for the first time was still a developing country; along the street from the airport to the hotel where the workshop was held, I saw some people sitting and chatting by the street. My impression was that Singapore is a nice warm place where people enjoy their lives when compared to Japan where everybody was busy or at least behaved like being busy. Singapore is a hot place, but we enjoyed a walk on Orchard Road. In 1990, KK then organized the *International Conference on High Energy Physics* (Rochester Conference) in Singapore. Prof. Yoshio Yamaguchi was a principal member of the Japanese delegation and I helped him to organize some sessions. Around that time, theoretical physicists were very much interested in the Rochester Conference.

In the mid 1990s, I visited National University of Singapore (NUS) as part of a scientific exchange program between NUS and the University of Tokyo. I started discussions on physics with Cho Hiap Oh, and I visited him often later, writing several papers together. On the occasion of my visits to NUS, KK often entertained me at nice restaurants. I also attended many of the international symposia and workshops organized by the Institute of Advanced Studies (IAS) of Nanyang Technological University directed by KK. I had the chance to see many old friends of mine on these occasions.

Now Singapore is a highly developed country in Asia and also strong with a worldwide presence economically, and the style of the operation of IAS by KK will be replaced by another that is not clearly seen yet. But I believe that the seed sowed by KK will develop further in the future together with World Scientific which provides a free forum to present unconventional or unorthodox ideas in various fields of science via scientific journals.

K.K. Phua's Contribution to International Physics Community and the International Union of Pure and Applied Physics

生
日
快
乐

Leong Chuan Kwek
IUPAP Deputy Secretary General

Bruce McKellar
IUPAP Past President

Michel Spiro
IUPAP President

First, we would like to wish our dear friend, KK, a happy 80th birthday. KK has distinguished himself as a physicist, a successful entrepreneur, a science promoter and a notable educator.

IUPAP in Singapore

When K.K. Phua realized that the International Union of Pure and Applied Physics (IUPAP) would want to move its office from London some time in 2010-2, he set out to find out more in 2010 about the involvement of the secretariat and what it means to establish the office in Singapore. At that time, he was appointed as the Founding Director of the Institute of Advanced Studies (IAS) at Nanyang Technological University (NTU), a marvellous institute which he established around 2005–2006.

One of us, Bruce McKellar, a long time friend and acquaintance of K.K. Phua, had just completed his 3-year term as a Vice-President at Large. Bruce was a frequent visitor to Singapore. Singapore lies roughly mid way between Europe and Australia, and Bruce had to travel often to Europe and the UK each year. KK talked with Bruce on several of these occasions and Bruce convince KK that the move to Singapore would be ideal, at least for the next three years. It was also made clear to KK that the move of the office to a new location could be in 2014, and that IUPAP needed considerable financial assistance from the host of its Office.

KK made a preliminary offer in 2012 and the issue was discussed earnestly at the IUPAP executive council. At that time, the NTU President, Bertil Andersson, and KK had several meetings regarding the matter with the then IUPAP President, Cecilia Jarlskog, as well as Bruce who was then the President Designate of IUPAP. These discussions quickly led to the official establishment of the IUPAP office in Singapore on 1 January 2015, with the offices supplied by NTU, and resources supporting the Office were provided by NTU and the Institute of Advanced Studies at NTU and by World Scientific, a publishing company founded by KK many years ago.

Activities in Singapore

Prior to the establishment of the Singapore Office, KK made sure that Singapore was actively involved with IUPAP: he featured IUPAP in many of the activities of the NTU-IAS, and increased the visibility of the new office worldwide. One such example is the 28th General Assembly of IUPAP which was held at the Nanyang Technological University in Singapore from 5–7 November 2014. This close connection with IUPAP in the activities and conferences at the IAS continued after the Office was established. Examples were the celebration of the 90th birthday of Freeman Dyson, and the 100th anniversary of the birth of General Relativity. Another example was the subsequent participation of Michel Spiro, when he was appointed as the President Designate of IUPAP, to the Inter-Continental Academia (ICA). The ICA hopes to create a global network of young future research leaders. Each session of the ICA is helmed by two Institutes of Higher Learning. The third

ICA in 2018–2019 was jointly organized by the NTU Institute of Advanced Studies (KK) and the University of Birmingham Institute of Advanced Studies. The topic focused on Laws, Dynamics and Rigidity. It was a great event connecting eminent scientists, historians, economists, social scientists and researchers with many carefully selected promising young researchers. KK was instrumental for facilitating the event.

Dedicated Staff

A dedicated staff member, Maitri Bobba, was hired by KK to work full time on the IUPAP business. Maitri Bobba, with the help of Sun Han from World Scientific, provided significant administrative support for the IUPAP's office in Singapore. She was efficient and she was full of energy and vitality. This was clearly KK's inspired, and wise, choice.

However, changes in the Directorship of the IAS, and in the management structure of NTU led the transfer of the IUPAP office out of NTU at the end of 2020.

With this special tribute, IUPAP wishes to deeply thank Prof. K.K. Phua, and expresses best wishes for this turning moment in his life.

An Enterprising Theoretical Physicist

生
日
快
乐

H. T. Nieh

Tsinghua University, P. R. China

Fifty years! In the fall of 1972, Phua Kok Khoo (潘國駒), when he was 30, came to the Institute for Theoretical Physics at the State University of New York at Stony Brook, an Institute that was established in 1966 with Professor Chen-Ning Yang (楊振寧) as director, for an extended visit. He and I have since become friends. My memory is somewhat vague now, but I remember he talked about the conditions in Singapore and the then struggling Nanyang University, where he was at the time a young faculty in the Physics Department.

The 1972 visit to Stony Brook was his first, which was followed by a few more visits in the 1970s and 1980s. During one of his visits in the 1970s, he talked about the high prices of science textbooks and expressed interest in finding ways to reduce the book prices for third-world countries. In one of the visits, I remember, he told me with excitement that his idea of making science books accessible to third-world readers had received enthusiastic support from renowned scientists like Abdus Salam, C. N. Yang, Steven Weinberg, etc. He was apparently pursuing the idea of starting a publishing business — not motivated by profit, but by serving the need of third-world science readers. World Scientific Publishing Co. was thus born!

It was much later in the 1980s, KK told me in one of his Stony Brook visits that World Scientific would, in addition to publishing science books, start publishing science journals. I was a little "shocked" by his ambition. During the early years of World Scientific, in my impression, most of its publications were reprint volumes of published physics papers with introductory chapters by expert physicists. I thought, publishing reprint volumes is one thing, but publishing new journals of original scientific research would be an entirely different, and much more difficult, matter. The authors, the subscriptions by institutions, etc., are all uncertain! I remember to have voiced my doubt out loud to him. Well, KK and his dear wife have proved me wrong. World Scientific has grown to be one of the leading scientific publishers in the world and the largest international scientific publisher in the Asia-Pacific region!

In the fall of 1989, Prof. Ngee-Pong Chang (章義朋) of the City University of New York initiated organizing an association of overseas Chinese physicists. Prof. Chi Yuan (袁旂), also of the City University of New York, Prof. Leroy Chang (張立綱), then at IBM Watson Lab, Prof. Jason Ho (何天倫), of Ohio State university, and I joined Ngee-Pong in a City University seminar room for the first organizing committee meeting. One of the questions discussed in the meeting had to do with funding that had to be raised to cover the expenses for such a non-profit organization. One of the funding avenues is donation by individuals. I remember, Ngee-Pong said he would talk to KK about it. Indeed, KK has been one of the major donors to the Overseas Chinese Physicists Organization (OCPA). To my knowledge as the initial OCPA Treasurer, KK contributed the funds for the various OCPA Prizes.

KK, to me, is a person who always strives to achieve things, and is willing to help others achieve as well, for a good and just cause.

KK, Happy 80th Birthday!

For the many encounters with KK over the years, I can only find a few 1993 pictures. In the fall of that year, KK and Mrs. Phua came to Stony Brook for a visit. I told them about the seminar series on China business and economy organized by Prof. Nien-Tzu Wang (王念祖) at Columbia University, and they expressed interest. We then jointly, a few days later, made a visit to

Columbia to meet with Prof. Wang. And, fortunately, a few pictures were taken and kept.

About KK in Singapore

C. H. Oh

Department of Physics,
National University of Singapore

Kok Khoo, popularly known as KK, is not only a famous physicist in the international community, he is also well known and respected in Singapore for his many community services, contributing and leading in several educational activities and programs in schools, junior colleges and universities. Indeed, KK has always been playing a leading role in raising Singapore as an international hub for science education and science research.

Let me cite three programs for which KK plays a pivoted role.

(I) Tan Kah Kee Foundation (TKKF) and Tan Kah Kee International Society

Apart from supporting world class conferences and co-organization of conferences with other academic institutions, a very successful annual event is the Tan Kah Kee Young Inventors' Award (TKKYIA). The aim of TKKYIA is to "encourage young students to be curious and apply scientific principles to create innovative solutions in inventions". Let me quote, "It is jointly organized by the Tan Kah Kee Foundation (TKKF) and the Ministry of Education with the support from Science Centre Singapore. Each year, the award ceremony is a grand event held at the Science Centre Singapore and attended by a Minister or senior government official from the academia or research institutes. Keenly supported by schools and research institutes, the event includes an exhibition of all award-winning entries."

(II) International Science Youth Forum (ISYF) at the Hwa Chong Institution

In 2009, KK initiated and organized the first ISYF with support from the Singapore Ministry of Education, Hwa Chong Institution (HCI) and Nanyang Technological University (Institute of Advanced Studies)

> "The overarching aim of the Forum is to create a high-profile platform for elite science students and teachers from Asia to network with their counterparts during the event. The Forum aims also to provide an avenue for intellectual dialogue and engagement with Nobel Laureates and other eminent scientists who will share their life experiences with the students. This will create greater awareness amongst the youth about the importance of broad-based knowledge, diverse interest, deep passion, tenacity and a global mindset in the pursuit of science excellence."

For the ISYF 2022, see http://isyf.hci.edu.sg/.

(III) GYSS (The Global Young Scientists Summit) program

KK initiated and started GYSS program in 2010 and together with the NRF (NATIONAL RESEARCH FOUNDATION, Prime Minister's Office, Singapore), the first GYSS was held in 2011. "The GYSS is a multi-disciplinary summit that provides a unique platform for young researchers and renowned scientists to discuss openly about technology trends, scientific discoveries and breakthrough innovations".

The program is still ongoing:

> "GYSS is an annual meeting of researchers and technology experts from various scientific disciplines from various countries."

See the Website for the 10th Global Young Scientists Summit 17–21 January 2022, Singapore.

These three programs undoubtedly have led to decisive contributions and lasting impact on the education and scientific community.

I first met KK at the *International Meeting on Frontiers of Physics* in Singapore in 1978, I was then working at the University of Science of Malaysia, Penang. He was so warm and generous and I felt that we had already met long time ago, a feeling probably shared by many people who have known KK. He played the key role in organizing this very successful meeting. It was probably the first large-scale international conference that was organized in Singapore and it brought Singapore onto the map of international physics. The *Proceedings, International Meeting on Frontiers of Physics*, was published in two volumes by the Singapore National Academy of Science. Our next meeting was in January 1980 when we both attended the 广州粒子物理理论讨论会, *1980 Guongzhou (Canton) Conference on Theoretical Particle Physics*, 5–10 January 1980, Guongzhou, China. After the conference, the organizer arranged trips to Beijing, Hanzhou and Shanghai. We got to know each other better, see the photo taken at the Great Wall in 1980.

With KK's help, I moved from Penang to work at the National University of Singapore (NUS) in 1983. I am indebted to him for introducing me to work at NUS in Singapore. Since then, we have been working together. He introduced me to his research area on particle phenomenology and multiple particle production, we have collaborated on several papers. In our joint research work and discussions, he was brilliant, very original and offering many ideas, almost instantly tackling problems from multiangles. In the contributed paper section, of this volume, we submitted a paper "Partition temperature model", which is in some sense a continuation of our joint paper done in 1988, *Phys. Rev. D* **38**, 1004 (1988).

As is well-known, KK initiates and promotes many conferences in Singapore, the list is too long, let me name just a few physics conferences organized in the last century:

1. *International Meeting on Frontier of Physics (invited talk)*, 14–18 August 1978.
2. *First Asia Pacific Conference in Physics*, June 1983.
3. *First Asia-Pacific Workshop on High Energy Physics*, 21–27 June 1987.
4. *The 25th International Rochester Conference on High Energy Physics*, 2–8 August 1990.

At the turn of the present millennium, KK started a new series of workshops/conferences on quantum information and computation, the first was organized with NUS Physics Department, *First Asia-Pacific Workshop on Quantum Information Sciences* (APWQIS_2001). This first workshop is responsible for the formation of quantum information technology group at NUS Physics Department, which eventually led to the formation of the Centre for Quantum Technology (CQT), one of the research centers of excellence in Singapore.

I am sure Kok Khoo will continue to actively contribute to the scientific and educational community for many years to come.

Biological Liquid Crystal Research and K.K. Phua's Visionary Support

生
日
快
乐

Zhong-can Ouyang

Institute of Theoretical Physics,
Chinese Academy of Sciences, Beijing 100190

In 2021, there were two major events in the Chinese scientific community: Chinese scholars all over the world held celebrations for the centennial birthday of Dr. C. N. Yang and for the 95-year birthday of Dr. T. D. Lee, both being the first Chinese Nobel laureates (in 1957) and world-renowned physicists. During these events, I was invited to write two commemorative articles, Dr. Yang's support for the interdisciplinary science of physics and biology in China (published officially by the Institute of Advanced Research, Tsinghua University, 2021-10-20), and Dr. Lee and statistical mechanics (published in *Modern Physics* (in Chinese), 2021, **33**(5): 40–50). These two events were heavily reported in media. Another great event, however, was rarely reported but also definitely worth remembering: the 40th anniversary of World Scientific Publishing Company (WSPC), the largest academic publishing company in the Asia Pacific region, which was founded by Prof. K.K. Phua. Since the establishment of WSPC, Prof. Phua has worked closely with top publishing houses and famous scholars in China for a long time, and is committed to introducing the academic achievements of Chinese scholars to the world. Therefore, he was the only Chinese who received recognition from 15 winners from 14 countries at the ceremony of the 15th China Book Special Contribution Award (Beijing, September 14, 2021). As the beneficiary of his great contributions, I would like to take this opportunity that celebrates Phua's 80th birthday to express my deepest respect for him.

The reason why I chose the title "Biological liquid crystal research and K.K. Phua's visionary support" for this article is inspired by the 2021 series of activities themed as "liquid crystal" which were organized by T. D. Lee Science and Art Lecture Fund of Shanghai Jiaotong University. The theme was selected and launched on March 21, 2021 by Professor Frank Wilczek, winner of the 2004 Nobel Prize in physics and director of T. D. Lee Institute. In the introductory article, *Finding the Magic in Liquid Crystals*, he wrote

"The uncanny material in your TV display may hold astonishing potential for creating future technologies.

The most glamorous fields in physics are probably cosmology and the search for new fundamental laws. While those subjects are glorious, there's also another frontier to relish: the pursuit of magic. Of course, here I don't mean trickery but magic in the spirit of the science-fiction writer Arthur C. Clarke's famous law: Any sufficiently advanced technology is indistinguishable from magic.

. . .

But there's another branch of modern magic, under-appreciated and still vastly underdeveloped, that has humbler origins. You can see a hint of it in the gooey residue left behind in your soap-dish. Elementary science classes advertise three "states of matter" — solid, liquid and gas — but that barely scratches the surface of what's out there. That goo is something else: a liquid crystal.

Liquid crystals, which flow like liquids but interact with light like crystal, are a distinct phase of matter. They are usually made from long organic molecules. Their basic secret is that those molecules are oriented in regular patterns, while their centers can move freely. The oriented molecules act like little antennas for the electromagnetic waves we see as light. They absorb those waves and then retransmit them in altered forms.

Because liquid crystals combine light-altering properties with fluidity, they can provide ultra-flexible, color-sensitive lenses and polarizers. This makes them tremendously useful for building visual displays. Indeed, liquid crystals are central players in most modern computer screens.

The mathematical theory of liquid crystals combines the intricate symmetry of crystalline patterns with the dynamical richness of liquid flows and then examines how these elements interact with light. In 1991, Pierre-Gilles de Gennes won the Nobel Prize in Physics for his contributions to that theory.

Yet when it comes to our understanding of liquid crystals, the best is yet to come. In fact, liquid crystals are central to life itself. A special kind of two-dimensional liquid crystal, closed up into sphere-like surfaces, forms the membranes that define the boundaries of cells and of functional units within cells. These crystals can selectively dissolve complex protein molecules, thus accommodating cellular eating, digestion, excretion and respiration. They can also grow, bud and fission — activities that are the soul of biological development and reproduction.

Human engineers haven't caught up to nature's skill in this medium. Our machines don't reproduce, develop, heal or regulate intercourse with their environment with anything approaching the sophistication of biology. A major bottleneck is that although we have good equations for liquid crystals, we're not yet good at using those equations to guide creative design.

This problem may be too complicated for unaided human brains. It seems likely that computer algorithms, which learn by doing (or, more accurately, by simulating), will yield more successful designs than conventional human logic. In this way, today's magic will conjure up tomorrow's."

In 2021, T. D. Lee Science and Art Forum invited me to give a talk entitled "From liquid crystal display to theory of liquid crystal biomembrane", which is the opening speech of the series of activities. This honor should be attributed to Prof. Phua and WSPC who promoted my research works to be much more visible internationally. As Prof. Wilczek pointed out, "liquid crystals are central to life itself. A special kind of two-dimensional liquid crystal, closed up into sphere like surfaces, forms the membranes that define the boundaries of cells and of functional units within cells". The first international monograph on biomembrane physics was written by us and published in 1999, which was totally encouraged and supported by Prof. Phua and WSPC. This book, entitled *Geometric Methods in the Elastic Theory of Membrane in Liquid Crystal Phases* (by Zhong-can Ou-Yang, Ji Xing, Liu, Yu-Zhang Xie, World Scientific, 1999), introduces the necessary liquid crystal physics and differential geometry methods to study biomembranes. It was highly praised by Japanese biologists immediately after its publication. In his book review, Prof. Ryota Morikawa from School of Life Science, Tokyo University of Pharmacy and Life Sciences, wrote

"Biological cells constituting small organs have a number of nm-thickness bio-membranes with the appropriate shapes for living. Therefore, studying the shape of the phospholipid bilayer film and its fluctuation property is very important for understanding things related to life phenomenon. In physics, theoretical and experimental studies on the shape of the bio-membranes have been conducted using methods based on the idea of elastic theory. For example, a blood cell with a form like a white ball pushed at the middle and closed phospholipid bilayer film, artificially made as a model of biological cell (vesicle), can change their shapes flexibly, depending on temperature, osmotic pressure, etc.

In 1973, Helfrich formalized the bending elasticity energy of the membrane film to analyze these shapes by using the variational method, and successfully showed a shape of stable pattern of vesicles. Afterwards, many researchers such as Peterson, Svetina, Lipowski, and Seifert have actively conducted the shape analysis of vesicles by using computer calculation. This book is based on physics of the liquid crystals and is a textbook presenting the elastic theory of the film. In the study of shapes of the film, using computer simulation is very popular owing to the nonlinearity of basic equations to be solved. However, the theoretical analysis is essential to support such computer calculation results, and the interest of the authors of this book is therein. This book describes the process for deriving the Frank elastic energy of the film in detail, on the basis of the continuum elastic theory of nematic liquid crystals, and shows a powerful way for analyzing the stability of the shape of the vesicles for a couple of symmetric cases. Generally, it is essential to develop the idea of the elastic energy of medium with curved surfaces for analyzing the shape of vesicles, meanwhile in the whole description of this book, the authors are always keeping in mind that bio-membranes are comprised of liquid crystal molecules of phospholipids. In particular, in the last chapter, they describe on the spiral structure films that appear due to the difference of the state of the lipid chain. There are many publications on the physics of the film, but they are conference records, collections and so on, in a very compact form. On the other hand, this book precisely describes the basis of differential geometry of curved surfaces. Accordingly, the derivation process of equations is easy to follow up. This book will be a valuable textbook for readers who are going to learn physics of membranes from now on. Nowadays, many ideas are presented related to life phenomenon from a view point of physics, and new scientific field called "artificial life" is growing up. Among them, physics of membranes uses an idea of classical physics related to elastic theory, but it can target a modeling of cells, i.e. biological basic element. Therefore, I believe it will gradually come into the limelight in science." (This comment is translated from the original Japanese version which was published in *Journal of the Japanese Physical Society*, 2000, **55**(4): 297.)

Perhaps it was this book review that prompted WSPC to list our book as one of the key publications of condensed matter physics in 2000. But what moved me more was a short message sent to me by Prof. Phua on April 2, 2019:

"Dear Ou-Yang
RE: GEOMETRIC METHODS IN ELASTIC THEORY OF MEMBRANES IN LIQUID CRYSTAL PHASES (2nd Edition)
I am pleased to inform you that your above book is doing well and we will continue to do active worldwide promotion. We are hoping to publish more books on topics in biophysics. Please feel free to recommend potential authors from China, US or India to us. Thank you for your support.
With best regards.
Yours sincerely
KK Phua"

WSPC is a major publisher of collected works of winners of Nobel Prize, Wolf Prize and Shaw Prize. It is also a leading publisher in physics, mathematics, life science, economics and other fields. In fact, it publishes more than 600 new books and more than 140 academic journals every year. I can imagine how busy Prof. Phua is, since he is the chairman of WSPC. Therefore, I was really surprised that he could remember our humble work on the little-known topic biological liquid crystal and generously show his appreciation, and I was also curious to know why. After receiving Phua's message, I searched online the quotation or comments of our book, *Geometric Methods in Elastic Theory of Membranes in Liquid Crystal Phases* (2nd ed., by Zhanchun Tu, Zhong-can Ou Yang, Jixing Liu and Yuzhang Xie, Singapore, World Scientific, 2017), and found that it has been recognized by more researchers. For example, John D. Clayton, a researcher at the US Army Research Laboratory and a professor at the University of Maryland, published a book review in the famous British magazine *Contemporary Physics* (2018, **59**: 431–433), as follows

"This book covers a number of general and specialised topics in differential geometry of surfaces as applied to membrane elasticity. Biological substances with properties of liquid crystal phases are the major application. However, many derivations and methods of analysis are general enough to be applied to other problems in mathematical physics. The stated purpose of the book is to deliver a comprehensive treatment of the conditions of mechanical equilibrium and elastic deformation of membranes viewed as a surface problem in differential geometry. Solutions that minimise the global free energy functional are intended to provide information on shape formation of membranes in their liquid crystalline state. **Besides the introductory material, most of the content is focused on intricate derivations of governing equations and their analytical outcomes. Much of the scope summarises or builds upon prior research by the authors who occupy scholarly positions in highly reputable Chinese academic institutions and hold exceptional publication records in this research field. Numerical and experimental methods are mentioned only in brief.**

This book, which as stated in the preface is the Second Edition of a work published some twenty years prior, is a compact volume of moderate length, 274 pages. The overall presentation is systematic. Basic physical motivation is given first, followed by mathematical preliminaries, meaning tools of differential geometry necessary to analyse complementary physical problems. Such tools are then specifically applied to treat different physically valid scenarios addressed in sections of each main chapter. Methods of classical as well as more modern geometry are used as is most appropriate for each application. Underlying topics in mathematical physics include geometry of curves and surfaces embedded in 3-space, the method of moving frames, exterior differential forms, and energy minimisation. Governing equations include kinematic and compatibility conditions as well as Euler–Lagrange equations of static equilibrium. The latter are most often derived by considering the first variation of a free energy functional for the elastic membrane. In some cases, stability and symmetry breaking are considered via use of the respective second and third variations of the energy functional. Dynamic problems, for example, those involving frequency response, waves, inertia and/or phase kinetics, are outside the scope of the text. The style is of moderate formality: derivations are mathematically rigorous but are not overly burdened with formal language (e.g. no highly formal mathematical theorems and proofs) that would make the work inaccessible to more applied researchers in branches of physical sciences, engineering, and other natural sciences outside the domain of pure mathematics.

The book contains six main chapters. Chapter 1 provides scientific background on liquid crystal bio-membranes, serving as physical motivation for more mathematically intensive treatments that follow in later chapters. History of the scientific discipline is discussed and basic definitions and classification schemes for liquid crystals, phases, biochemical properties, and vesicles are discussed. The remaining chapters tend to be organised as follows: geometry and kinematics, a free energy functional depending on geometric variables, equilibrium equations obtained by the first variation, and then potential solutions and their analysis. Chapter 2 deals with curvature elasticity of fluid elastic membranes, including classical geometric identities and free energy functions of F. C. Frank and W. Helfrich for liquid crystals. Chapter 3, the longest chapter, covers the shape equation and its solutions for lipid vesicles. A number of particular possibilities, for example, the sphere, Clifford torus, and Dupin cyclide, are addressed. Open lipid membranes are considered in Chapter 4, which features geometric analysis using the method of moving frames. Chapter 5 covers tilted chiral lipid bilayers, including general theory and solutions to governing equations for bodies of various shapes without free edges. Advanced topics are introduced, but are not thoroughly analysed, in Chapter 6: nonlocal theory, numerical simulations using the Surface Evolver software, phase field methods, and composite shell models. This final chapter ends with a terse extension of membrane theory to describe low-dimensional carbon materials such as graphene and carbon nanotubes. Appendices provide brief background information on tensor calculus in N-dimensional space, the gradient operator for smooth surfaces, derivation of the static equilibrium conditions from a force balance rather than energy variation, and lastly, surface tension in a lipid bilayer.

The topics addressed in each main chapter are described in sufficient detail and mathematical rigour to provide insight into each and to satisfy the overall objective of the text. Somewhat conspicuously, the book's main content jumps from qualitative definitions and examples from biological and chemical physics in Chapter 1 to rather advanced mathematics of geometry and energy minimization in later Chapters, without much transitional discussion of science that would fall in between. For example, fundamental discussion of the physical origins of representative free energy functions [e.g. eqns. (2.97), (2.104), and (3.38)] from the perspective of condensed matter physics [1] is mostly excluded in this regard. Membrane theory is used from the outset without justification. The book might benefit from an explanation of why two-dimensional membrane theory with curvilinear coordinates is advantageous over a three-dimensional Cartesian formulation for liquid crystal phases.

Though nanometer-scale phenomena are modelled, the book does not consider statistical physics nor quantum mechanical origins of material properties, for example, as in [1,2]. Perhaps in future editions, hydrodynamics and solid mechanics (e.g. minimally three-dimensional elasticity theory for solid and liquid crystals [1,3]) could be briefly incorporated, demonstrating how elastic membrane theories can be contrasted with or extracted from fully three-dimensional theories. A more thorough treatment of finite deformation kinematics and differential geometry pertinent to field theories of continuum mechanics [4], enhancing introductory material discussed in their Appendix A on tensor calculus, would also be a useful supplement. Analysis of elastic liquid crystal membranes containing topological defects such as dislocations and disclinations [1,3] would also be interesting avenues of future research for methods developed in this text.

The intended audience of this book is advanced graduate students and researchers in applied mathematics and mathematical physics. The monograph is self-contained. However, many topical descriptions may be too brief for students and researchers without prior training, for example, knowledge up to the level of a second-year graduate student in continuum mechanics, surface elasticity, and differential geometry. Regardless, it does serve as a highly useful and systematic reference work for moderate to more advanced practitioners of differential-geometric analysis of deformable surfaces. The book contains numerous examples of derivations of governing equations for elastic membranes in curvilinear coordinates and examination of corresponding general solutions. Some boundary value problems are worked through in the book, but no formal problem sets found in course textbooks are given. The book is thus deemed more suitable as a research reference than a textbook to be used for graduate-level instruction.

Each chapter contains its own bibliography. Though usually not exceptionally long, these reference lists are sufficient to address main topics in corresponding chapters, and they do often contain more thorough references on specific topics that adequately supplement the text. The authors are careful to cite particular works credited with origins of various theories or mathematical results first derived elsewhere. The quality of the editing, printing, and binding are excellent, meeting the usually high standards of recent World Scientific books. The book contains numerous line figures in black and white that accompany the aforementioned examples. These are all rendered with precision and quality, nicely complementing the careful derivations in the text. Grayscale contours of predicted shapes of membrane structures are also found in several chapters that provide exceptional insight into equations describing their properties. Chapter 1 also contains instructive grayscale illustrations that accompany fundamental definitions from chemistry and physics, many similar in style to those in [1]. Equations are superbly formatted, and the font sizes are appropriate, for example, with subscripts and superscripts large enough to be clearly legible throughout the book. The book contains a detailed Table of Contents and a brief but sufficient subject index. No author index, glossary, nor list of symbols are included.

In summary, the book *Geometric Methods in Elastic Theory of Membranes in Liquid Crystal Phases* is highly recommended as a reference for advanced graduate students and scholars involved in geometric analysis of membranes and other elastic surfaces. Valuable techniques may be learned from the book's model constructions and sequential derivations and presentations of governing equations. Detailed analysis and solutions enable the reader with an increased understanding of the physical characteristics of membranes in liquid crystal phases such as their preferred shapes."

The above review lists four reference books:

[1] **Chaikin PM, Lubensky TC.** *Principles of condensed matter physics.* **Cambridge: Cambridge University Press, 1995;** [2] **Wallace DC.** *Statistical physics of crystals and liquids: a guide to highly accurate equations of state.* **Singapore: World Scientific, 2003;** [3] **Clayton JD.** *Nonlinear mechanics of crystals.* **Dordrecht: Springer, 2011;** [4] **Clayton JD.** *Differential geometry and kinematics of continua.* **Singapore: World Scientific, 2014.**

Among these famous monographs in condensed matter physics, [2] and [4] were published by WSPC, and our book also belongs to (soft) condensed matter physics. Hence, I think Prof. Phua's foresight on the trends of condensed matter physics was the reason that he provided strong support for our book.

There's another interesting thing about the book worth to be mentioned, i.e. the 1st edition of the book helped me "win" the debate with Prof. T. Lubensky (Department of Physics and Astronomy of the University of

Pennsylvania), the second author of the above reference book [1], at Isaac Newton Institute for Mathematical Sciences, Cambridge University. In 2013, I was invited to participate in the advanced seminar on "liquid crystal mathematics" led by Prof. Lubensky. On June 6, I did a seminar entitled "Geometric elasticity theory of liquid membrane bubble: Helfrich model of lipid bilayer and its application in soft matter", in which I showed our theory can give an accurate analytical solution of the inclination angle of cylindrical spiral bands only by using the elastic energy of cholesteric liquid crystal (Komura, Ou-Yang, *PRL* **81** (1998) 473)

$$\phi_0 = \arctan\left[\left\{\frac{8}{3}\cos\left(\frac{1}{3}\arccos\frac{5}{32}+\frac{1}{3}\right)\right\}^{1/4}\right] = 52.1°$$

This is in perfect agreement with the experimental observations on the model bile membrane by researchers from the Harvard Medical School and Department of Physics of MIT (D. S. Chung *et al.*, *PNAS* **90** (1993) 11341)

$$\phi_0 = 53.7 \pm 0.8°$$

Prof. Lubensky (Lubensky T. C., Prost J., *J. Phys. II* (France) **2** (1992) 371) and the Nobel prize laureate P. G. de Gennes (*C.R. Acad. Sci. Paris* **304** (1987) 259) also worked on this problem. They used a ferroelectric liquid crystal model, which assumes that either side of a ferroelectric ribbon has positive or negative charge. When the ribbon is about to be coiled up, the internal repulsion due to like charges will lead to an open spiral ribbon. Therefore, they deeply questioned our explanations based only on elastic energy. As the chief reviewer of my candidacy for member of the Third World Academy of Sciences (TWAS) in 2003, de Gennes wrote in his letter to TWAS that he highly appreciated my theoretical works on biomembrane and carbon nanotube, but did not like my work on spiral membrane (at the end of the letter, however, as a great scientist, he still expressed generously his strong recommendation for me). In Newton Institute, it was the first time that Prof. Lubensky and I could have a face-to-face debate on that topic.

After my talk, he asked me a very sharp question: "if we only consider the elastic energy, isn't the cylindrical tube formed by cholesteric liquid crystal the lowest-energy state? Why does it crack and form an open spiral?" I gave a brief qualitative answer: In the elastic energy density of cholesteric liquid crystals, there is a linear curl term of the orientation vector field, so the corresponding area integral will become a line integral along the boundary; If the elastic coefficient is negative, the belt will choose the appropriate spiral chirality to ensure that the total energy of the negative line integral is always lower than that of the cylindrical tube, so the latter is not the lowest-energy state. Prof. Lubensky then asked if I could show him the detailed calculation. Because this calculation is very complex and cannot be explained clearly in a few words, I felt embarrassed at first, but I suddenly thought of my book which gives all the necessary details to understand our theory and can help me respond to his request. This book was not at hand, but have been available here in the institute. In the spring of 2004, I was invited to Newton Institute for the first time to participate in the advanced seminar on "Statistical mechanics of molecular and cellular biological systems". Newton Institute is the world-famous center to promote interdisciplinary researches (e.g. the great mathematician A.Wiles announced the proof of Fermat's theorem there on June 23, 1993). Among the "classmates" of that seminar were J. Walker (the winner of the 1997 Nobel Prize in chemistry), S. Edwards (a top polymer physicist and Cavendish Laboratory professor). I remembered that the organizer had encouraged us to donate our monographs to the library of Newton Institute. At that time I donated the first version of my book (WSPC, 1999) , but I was not sure if it was still there after 9 years. Therefore, I went to the library trying to find the book to respond to Prof. Lubensky. Fortunately, the book had been kept perfect in the library and had the official retrieval number: in lib of ini: qc173ouy. It's this book that made us become good friends. Prof. Lubensky, as well as three other members of National Academy of Sciences of US (David Weitz, Steve Granick and Robert Austin), were then invited by me to give plenary talks on the 11th Conference on soft matter and living matter physics in China (Chongqing, November 9, 2018) . We were happy to have many more communications at that conference, and shared more ideas with more than 400 researchers and students from universities and research institutes across the country to attend the meeting.

K.K. Phua's attention to biophysics is not limited to biological liquid crystals. On February 22, 2017, he sent me the following email,

"Dear Ouyang:
We recently met Prof. Xiang-yang Liu. WSPC has a small English journal *Biological Reviews and Letters*. Liu suggested we should cooperate with you. You are the editor-in-chief of *Journal of Physics* (in Chinese), the articles can be translated into English. I hope you can join us as one of the editors of this English journal. This journal focuses on review articles. We recently invited Prof. Matthew Fisher to give a nice speech entitled "Are we quantum computers, or merely clever robots", which will be published in *International Journal of Modern Physics B*. Thank you!
Good luck!
KK Phua"

After I replied to the invitation, he immediately wrote to me,

"Thank you very much for agreeing to be one of the editors of *Biological Reviews and Letters*. This journal was proposed by Prof. Kerson Huang during his lifetime. Biophysics has developed rapidly and is closely related to life sciences. If you know any Asian (or non-Asian) scholars working on biophysics in Europe or U.S., you can recommend one or two suitable candidates, and we will invite them to be members of our journal editorial board."

Prof. Kerson Huang, who died in September 2016, was a famous statistical physicist and poet in the Chinese physics community. Huang's transfer to biophysical research began when he was employed as a lecturer professor (2004–2008) of Zhou Peiyuan Center for Applied Mathematics of Tsinghua University, which was founded by Prof. Chia-Chiao Lin, an internationally famous mathematician. Huang visited the center for two months every year and cooperated with Lin's group to study protein folding. After the experience in Tsinghua University, Huang was then employed (2009–2011) by Nanyang Technological University, and led a group in the Institute of Advanced Studies under the Director Prof. Phua to continue to study protein folding.

Prof. C. C. Lin was appointed as Distinguished Professor by Tsinghua University in 2002 (about June). At the seminar entitled "Applied Mathematics in the 21st Century", Lin declaimed that the problems of life science should become an important research field of applied mathematics. Because of the rapid development of experimental methods, people have accumulated a large number of experimental data, which make it possible to quantitatively study complex life phenomena. However, there is a lack of basic principles similar to physics laws in life science. Therefore, it is a very challenging problem to explain these experimental phenomena quantitatively, which also provides rich research topics for applied mathematics.

In order for the approach to life sciences, at the inaugural ceremony of Zhou Peiyuan Center and the first International Symposium on Frontiers of Applied Mathematics on August 30, 2002, Prof. Lin, as the president of the conference, invited 12 speakers including me, namely, David Benney (Professor, Department of Mathematics, MIT); Yuan-Cheng Fung (Member of National Academy of Sciences, UC San Diego, Bioengineering); Nancy Kopell (Member of National Academy of Sciences, Boston University, Department of Mathematics, Center for Biodynamics); Zhong-can Ouyang (Academician of Chinese Academy of Sciences, Institute of Theoretical Physics, Chinese Academy of Sciences); Da-Qian Li (Academician of Chinese Academy of Sciences, Institute of Mathematics, Fudan University); Lee A. Segel (Professor, Weizmann Institute of Science, Applied Mathematics, Israel); Frank H. Shu (Professor, President of Tsinghua University, Hsinchu, Taiwan); Theodore Y. Wu (Professor, Caltech, Engineering Science); Chi Yuan (Professor, Institute of Astronomy, Academia Sinica, Taiwan); Gong-qing Zhang (Academician of Chinese Academy of Sciences, School of Mathematics, Peking University); Zhe-min Zheng (Academician of Chinese Academy of Sciences, Institute of Mechanics, Chinese Academy of Sciences); Heng Zhou (Academician of Chinese Academy of Sciences, Department of Mechanics, Tianjin University).

In my talk, I reported some theoretical achievements in the research on biological liquid crystal. Among them, the analytical solution of the double concave shape of red blood cells, which was being investigated in biomechanics for a long time, can be derived from the shape equation of liquid crystal membrane, as below

$$\Delta P = 0, \quad \lambda = 0, \quad \sin \Psi(\rho) = C_0 \rho \ln(\rho/\rho_B)$$

$$C_0 < 0 \rightarrow \text{bioconcave}, \quad z = z_0 + \int_0^\rho \tan \Psi(\rho')d\rho'$$

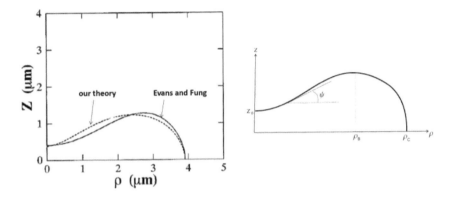

It is in good agreement with the experimental measurements of Prof. Yuan Cheng Fung's group in the 1970s (A. Evans, Y. C. Fung, *MICROVASC. Res.* **4** (1972) 335). Prof. Fung was deeply excited by my talk, he said, "Having your talk is like drinking excellent wine!" Shortly after my talk, Prof. Lin asked me to come to his office in Tsinghua, talking about his research plan. He believed that there was a deep similarity between the dynamics of protein folding and turbulence formation. He suggested that the theory and methods of the latter can be employed to study protein folding, and invited me to join the academic committee of the center and help his research group recruit postdocs. Unfortunately, shortly after Lin's death on January 13, 2013, this group was dissolved, and researchers from the Department of Applied Mathematics of Tsinghua returned to the department to continue teaching and scientific research, while the postdocs had to seek their own way out. I could only help few of them to find positions in some universities in China. What a pity that Tsinghua University, which has abundant research funds, could not realize Lin's long-term goal to study life science problems from the perspective of applied mathematics! In contrast, it's really admirable that Prof. Phua was so decisive to hire Prof. K. Huang after he left Tsinghua

University for Nanyang Technological University to continue research on protein folding, and found the journal *Biological Reviews and Letters* as proposed by Huang when he was alive.

Prof. W. Helfrich proposed the concept of spontaneous curvature of membrane and the corresponding elastic free energy of liquid crystal in 1973 (W. Helfrich, *Z. Naturuforsch*, **28c** (1973) 693), but it was not until 1987 that he and I derived the general equation of bubble shape with spontaneous curvature from rigorous surface differential geometry (Z. C. Ouyang, W. Helfrich, *PRL*, **59** (1987) 2486; *PRA*, **39** (1989) 5280). In addition to its application to phospholipid biomembrane, the Swedish chemist Prof. J. C. Eriksson and other scholars called this equation Generalized Laplace Equation for studying oil, water and surfactants **(S. Ljunggren, J. C. Erilkssen (1992); S. Komura and K. Seki (1993))**. The beauty of this equation attracted the attention of Prof. J. C. C. Nitsche whose research field was differential geometry. He pointed out in the preface to the revised version of his encyclopedia *Minimal Surface* (1993) that Helfrich's theory on fluid membrane is a major update to Poisson's elastic shell theory (1812). In 1990, based on this equation, we predicted the existence of a membrane torus with a radius ratio of root 2 (Z. C. Ou Yang, *PRE*, **41** (1990) 4517), i.e. the famous Clifford torus in mathematics. This was soon confirmed by experiments. This beautiful result helped me win the first Asian Outstanding Achievement Award of the Overseas Chinese Physics Society in 1993. However, what could really attract mathematicians' attention to the research of biological liquid crystals is due to the visionary support of Prof. Phua.

In 2013, K.K. Phua sent a letter inviting me to Singapore to attend a special birthday party for Freeman Dyson, a 90-year-old British American mathematical physicist and professor of Institute for Advanced Study in Princeton. Many great physicists (including several Nobel laureates) and his fans from all over the world attended the birthday party. Although Dyson did not win the Nobel Prize, he had a very high reputation in physics and popular science, and was respected by everyone. During the four-day celebrations (August 26–29), K.K. Phua arranged a 45-minute invitation talk for me, entitled "Overview of the study of complex shapes of fluid membranes, the Helfrich model and new applications". This talk may not have been very attractive to high-energy physicists or condensed matter physicists, but it deeply attracted Prof. Yang Chen (Department of Mathematics of the University of Macao). He also gave an invited talk, entitled "Determinant of Toeplitz + Hankel matrices, non-linear difference equations, and Painlev VI". Chen did his Masters under the supervision of Prof. Frederick Ernst who discovered with Isidore Hauser that under stationary and axial symmetry (or the situation with two commuting Killing vectors), Einstein's equation could be reduced to a nonlinear second order partial differential equation satisfied by a complex potential known as the Ernst potential incorporating the Maxwell Field. He was very interested in the third-order nonlinear differential equation of the axis-symmetric biomembrane bubble I derived, and thought it was likely to be transformed into a second-order nonlinear Painlev VI equation. Therefore, after the Singapore meeting, he invited me to visit University of Macao for collaboration. Although we had not yet made significant progress, he gave me an amazing surprise, i.e. the chance to introduce our research on biological liquid crystals to the international mathematics community. On his recommendation, I was invited to make a one-hour plenary speech at the 10th ISAAC (International Society of Analysis, its Applications and Computation, which is held once every three years) at University of Macao on August 3–8, 2015. More than 300 mathematicians attended the meeting. Among the 12 invited speakers were two famous Chinese mathematicians, Prof. Yum-Tong Siu of Harvard University and Prof. Ngaiming MOK of the University of Hong Kong (elected as academician of Chinese Academy of Sciences in November 2015). All the invited talks were then collected and published in *Springer Proceedings in Mathematics & Statiscs* (SPMS), one of Springer's most famous series, including our contribution (Zhong-can Ou-Yang and Zhan-Chun Tu, *The Study of Complex Shapes of Fluid Membranes, the Helfrich Functional and New Applications, SPMS* **177**, Tao Qian, Luig G. Rodino (Editors), *Mathematics Analysis, Probability and Applications — Plenary Lectures*, ISSA2015, Macau, China).

At the end of this article, I would like to express the heartiest gratitude and congratulations to Prof. Phua on his 80th birthday, who is more than a friend and teacher to me. I was lucky to meet Phua for the first time in 1990s when I helped Prof. Bai-lin Hao (academician of CAS) raise funds for the 19th International Conference on Statistical Physics held at Xiamen University in 1995. Phua suggested that we ask for financial support from Tan Kah Kee International Society. So Prof. Hao tried to meet Mr. Tan Keong Choon (the managing director of the society) and Mr. Chan Mo Pok (the treasurer of the society) in 1992, who were attending the preparatory meeting of JiMei University (at Xiamen, FuJian Province) at that time, to explain our budget and request. Mr. Chan Mo Pok promised to Prof. Hao that the society would give full support of $70000, an equal support to the International Conference on Electrochemistry held at Xiamen University in 1995.

Mr. Tan Keong Choon and Mr. Chan Mo Pok were very close relatives and friends of my wife Ms. Jin-hua Zhou. In fact, we call them uncles. Prof. Yan-hang Chen from Navigation College of JiMei University described the relationship between us and Tan Keong Choon in an article "The stories between JiMei University and physicists including Dr. C. N. Yang" (published in *Journal of the Alumni Association of JiMei Navigation College*), "Zhong-can Ouyang, a young academician of Chinese Academy of Sciences, has a deeper relationship with JiMei University. Let's start with his wife Ms. Jin-hua Zhou. When Zhou came alone to JiMei Navigation College, she was appointed to work in the library and live in Kerang Building on campus. We happened to be neighbors. Her husband Zhong-can Ouyang came back to visit her and told me that he was an alumnus of Tsinghua University and was pursuing the doctoral degree in theoretical physics. After his doctoral degree, Ouyang got the chance to go to Institute of Theoretical Physics of Chinese Academy of Sciences to do postdoctoral research, and Zhou then had to follow him to go to Beijing. However, they were still much concerned about the research in JiMei Navigation College. Ouyang had once developed a management software systems for a hotel, so he contacted the hotel to donate the software to the College to promote the scientific research. Zhou donated the printing/editing software systems she used in Institute of Theoretical Physics (including 16 floppy disks and texts) to the College to facilitate journal publications. At that time, this donation was extremely valuable. I remember it was brought to China from abroad by Chinese Academy of Sciences.[a] In 1990, entrusted by Tan Kah Kee Foundation, I went to the back street of Chongfu Temple in Quan Zhou city to visit Ms. Mo-yin Zhang, the nominally adopted daughter of Mr. Jing-xian Chen.[b] Her biological father Wu-wo Zhang,[c] from Hanchuan, Hubei Province, was once a member of the Chinese Revolutionary League responsible for Wuchang Uprising. Wu-wo Zhang was first hired by Jing-xian Chen as a tutor, and then they became sworn brothers. Jin-hua Zhou is the daughter of Mo-yin Zhang. When Jing-xian Chen's wife came to China from overseas, she always went to Quan Zhou to visit her adopted daughter. Mr. Tan Keong Choon[d] also often visited Mo-yin in Quan Zhou when he came to China. I never heard Jin-hua Zhou talk about this. In fact, Ouyang couple not only have a close working relationship to JiMei University, but also have a much deeper family relationship with the co-founder.[e]"

Our relationship to Mr. Chan Mo Pok can date back to the friendship between my father-in-law's family and Chan's family. Chan had many brothers. In the Kuomintang ruling period in Mainland China, in order to avoid the forced military service, he and his brothers tried to hide in my father-in-law's home. Therefore, every time when the Tan Kah Kee Award (the predecessor of the Tan Kah Kee Science Awards) was announced in XiYuan Hotel in Beijing, Chan Mo Pok would invite me and my wife to have dinner together in the Hotel.

Although Mr. Tan Keong Choon and Mr. Chan Mo Pok were business men, they were very generous and supportive to scientific research. As Prof. Bai-lin Hao mentioned in the article entitled "Mr. Chan Mo Pok and International Conference on Statistical Physics" (Fu Ji Yin Xiao Lu (in Chinese), by Hao Bolin, BaFang Culture, 2009), "On the eve of the International Conference on Statistical Physics, Mr. Chan Mo Pok also arrived in Xiamen in advance to learn about the progress of the preparations, together with me. During the conference period, he has been in Xiamen and arranged staff

---

[a]In fact, by Prof. Bai-lin Hao.
[b]The younger brother of Tan Kah Kee.
[c]Also known as Su-wu Zhang.
[d]The son of Jing-xian Chen couple.
[e]It means Mr. Jing-xian Chen.

in Hong Kong to help participants passing by, especially those who came to Xiamen by boat from Hong Kong … Almost every time when Zhong-can Ouyang and other colleagues and I traveled to and from Taiwan, we were warmly hosted by Mr. Chan in Hong Kong.[f]"

Here, in particular, I would like to thank K.K. Phua for his generous invitation to me to attend the International Conference on Prof. C. N. Yang's 85th birthday held in Singapore in November 2007. On the last day of the conference, Phua helped me and my nephew Mr. Ting Xu (who was studying in the Department of Biology of Nanyang Technological University) contact and visit Uncle Tan Keong Choon and have dinner with him. In his house, we saw many Chinese leaders' inscriptions on carrying forward Tan Kah Kee's educational ideas. As Phua pointed out in his article "the cooperation between Singapore and China in education is imperative" 10 years later (in March 2017), "Tan Kah Kee International Society inherits the "Kah Kee spirit" of recognition of the significance of education. JiMei University is of course the first choice for the Society for cooperation with China on education. I remember that it was Tan Keong Choon (former president of the Chinese General Chamber of Commerce) who first proposed to jointly set up a school of business administration at JiMei University, that is, jointly organized by Xiamen municipal government, JiMei University and Tan Kah Kee International Society in Singapore. In 1995, the School of Business Administration of JiMei University was officially established. During that time, Jin-ping Xi, the present Chairman of PRC, was very concerned about the development of JiMei University and the School of business administration." Because of the precise and keen decision-making of K.K. Phua, the current chairman of the Tan Kah Kee Foundation, the Kah Kee Spirit will continue to promote the vigorous cooperation between China and Singapore on education. On October 17, 2020, the list of members of the Sixth Council of JiMei University was officially announced. As alumni of JiMei University, my wife and I sincerely congratulate K.K. Phua on being elected as executive director again!

---

[f]It's before direct flights between the mainland and Taiwan.

K.K. Phua and ICTP

生
日
快
乐

K. R. Sreenivasan

New York University, USA

I am pleased to write this brief note about Professor K.K. Phua, with whom I got connected via the International Center for Theoretical Physics (ICTP) in Trieste, during my time as its Director. The Founding Director of ICTP, Nobel Laureate Professor Abdus Salam, was one of the inspirations for Phua and World Scientific. Thus there had been strong ties between ICTP and World Scientific during Salam's time, which I was glad to revive. This short article is built around this connection, with a few background remarks added.

Phua earned his Ph.D. in 1970 from the University of Birmingham, where he was trained as a theoretical physicist in the phenomenology of high energy collisions. By the time he completed his formal education, Singapore had already become independent from Malaysia, on its way to prosperity. But it had not yet reached its present state of affluence; for example, the per capita GDP at the time was about a fifth of that of the US, whereas the two are comparable today. But it was clear that the country was moving rapidly in higher education (among other areas). The Science Council as well as the Science Academy were established quickly, and university education was being invigorated. The new country offered many possibilities for an enterprising person like Phua; for instance, he was able to create the Institute of Advanced Studies at Nanyang Technological University and serve as its founding director.

But it was scientific publishing that most captured Phua's attention and energy. Until 1981, when Phua and his wife established World Scientific, there was no international scientific publishing house based entirely in Asia. World Scientific started on a small scale with five employees. Today, it employs about 200 people at its headquarters in Singapore and about 450 globally. It has offices in New Jersey, London, Munich, Geneva, Tokyo, Hong Kong, Taipei, Beijing, Shanghai, Tianjin and Chennai. Within about two decades from inception, it had established itself as one of the leading academic publishers, and, having published more than 12,000 titles by now, it is a large international scientific publisher in the Asia-Pacific region.

There are many reasons why World Scientific has become a success. The broad commitment of Asian and international scientific communities, among whom are several leading scientists, quality service provided from production to marketing, exclusive publication agreements with institutions such as the Nobel Foundation, competent staff with good work ethic are all contributing factors — but there is no doubt that the charisma, commitment and acumen of Phua were the key. He combined business savviness with love for science, particularly physics and mathematics. He kept up with important developments in science, emerging areas of research, as well as promising scientists.

Among the people who inspired and supported the publishing goals of Phua was Abdus Salam. It is unclear when exactly they first met (Phua was an undergraduate in Imperial College where Salam remained a professor even after he founded ICTP), but it is clear that the creation and mission of ICTP in service of scientists from developing countries had an impact on Phua. The two had a warm personal relationship, and ICTP for many years published under favorable terms its numerous conference proceedings with World Scientific.

Phua was a visiting scientist at ICTP. The records show that he was a visitor in 1972, perhaps later as well; it is hard to reconstruct records completely from so long ago. What is clear is that he wrote at least two papers, one under ICTP affiliation (*Lettere al Nuovo Cimento* **5**, 575, 1972) and another with an ICTP affiliate (*Phys. Rev. D* **19**, 1512, 1979); see Box 1.

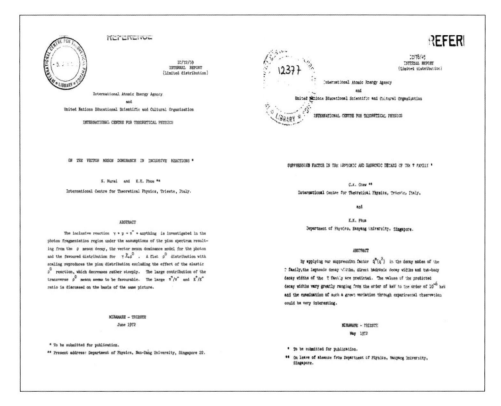

Box 1. The first page of two reports by Phua, from ICTP archives. These reports were later published in journals. See text.

Late in 2003, an ICTP associate from Congo-Brazzaville (Republic of Congo) visited me at ICTP and asked if I could buy a few books for the Physics Library at his university. He said that the library had been one of the casualties of the civil war of the late 90s. The peace accord had been signed only recently and they were in the process of rebuilding their physics graduate education with a master's degree. Somewhat shaken, but greatly admiring of his optimism and intention, I asked him how many books the library had at that time, to which he responded "something like 30" — all of which, by the way, had been bought by ICTP previously.

Buying him 30 more books would not have been a problem, but it would clearly have made no dent in the basic situation. It was then that I wrote for help to the US National Academies, American Physical Society, Springer-Verlag, Institute of Physics, and World Scientific. (I wrote to several others

---

Book Donations

The book donation programme distributes books, either received as donations or bought for this purpose, to libraries or individuals in developing countries.

A large donation of books was received from World Scientific (WSPC) in 2005 and 12,000 of them were distributed the same year. In 2006, the remaining 10,000 volumes were sent to requesting libraries or offered to visitors to take away personally.

Numerous requests for new books were received from scientists and Junior Associates. These covered all subjects, not only mathematics and physics, but also medicine and technology. Although funds were limited, thanks to special conditions agreed upon with a publisher and affordable prices offered by a bookdealer, the Library was able to purchase 224 new books for needy libraries as well as 750 textbooks and 6 memberships for 145 Junior Associates.

Box 2.   2006 Annual Newsletter, Marie Curie Library, ICTP, p. 150

as well — and visited two — but they were totally unresponsive even when the suggested participation was minimal, but that is a topic for a separate discussion.) I received a prompt note from Phua stating that he would send us recently published books if we paid for their shipping. In all we received 23,000 books from him (see Box 2). I had the privilege of distributing them (and other services such as free online access to journal papers) to a few libraries in great need of science and mathematics books: many libraries in the so-called least developed countries had stopped buying books for their libraries for long periods of time.

When my wife and I visited Singapore some years later, Phua and his wife, along with a few staff members, graciously hosted a lunch for us. When they also showed us around, we learnt more about Phua and World Scientific. We appreciated that opportunity very much.

Somewhere in his musings, Phua says that one should know how to interact with different levels of people, but being humble is an important common ingredient. I think he was giving expression to his own example. It remains for me to wish him and his wife many more years together in their adventures. Many happy returns, Dr. Phua!

Acknowledgment

I thank the staff of ICTP's Marie Curie Library, particularly its former librarian Mrs. Maria Fasanella, for their help in unearthing some data.

Challenging the Giants of Scientific Publishing!

生
日
快
乐

Bernard T. G. Tan

Department of Physics,
National University of Singapore

I have the privilege of being a good friend of Prof. K.K. Phua for 50 years or more. Though we were initially in two different Physics departments, he at Nanyang University and I at the University of Singapore, the faculty in the two departments had excellent relationships with each other. Indeed, they joined hands in the 1970s to form The Institute of Physics Singapore, and KK and I were very much a part of this process.

When the two universities merged in 1980 to form the National University of Singapore, we became colleagues. I had always admired his wide knowledge of fundamental physics, and always found him a very congenial and helpful colleague. A good example of his readiness to help others was his taking the trouble to carry back a friend's PhD thesis all the way back from the UK, to persuade him to make minor amendments so that the examiners could pass the thesis.

I also greatly admired his courage and foresight in the daring move which he made as a young academic to start up a new scientific publishing enterprise. This took a lot of guts as he was entering a field dominated by large internationally renowned publishing houses in the West. They certainly did not welcome an upstart publishing rival from Singapore, not then renowned for this area of publishing.

Within a decade, he confounded all the naysayers by quickly establishing World Scientific as a force to be reckoned with in the scientific publishing world. His great energy, quiet diplomatic approach plus his incomparable list of contacts with renowned scientists ensured that the venture would succeed. His tireless promotion of World Scientific amongst the myriad of renowned scientists who were his good friends (including many Nobel Laureates such as Abdus Salam and C. N. Yang) gave World Scientific the kind of credibility in the international scientific community enjoyed only by a select few.

His contributions to science extend far beyond scientific publishing, and include the promotion of scholarship and research, especially at the Centre for Advanced Studies at Nanyang Technological University where he was the founding Director. He did much for the promotion of science amongst the general public and young people and I was privileged to work alongside him when we started a series of public lectures on the theme of "Creativity in the Arts, Science and Technology". His enthusiasm for the promotion of physics education bore fruit when at his initiative, World Scientific started a journal devoted solely to this area, "The Physics Educator".

As he celebrates his 80th birthday, he and his wife Doreen, whose managerial judgement and sagacity also contributed critically to the success of World Scientific, can be justly proud of their achievements. World Scientific has indeed placed Singapore in a significant position on the world scientific stage, helping Singapore to cement its position as an international science and technology hub.

I and all of KK's friends and colleagues stand up to applaud him for a lifetime of achievements in science and science publishing, and know that he can rest assured that his work and his contributions in the field of science and science publishing will be long remembered and lauded. Bravo, KK!

For the Birthday Volume of K.K. Phua

生
日
快
乐

Jean Tran Thanh Van
*Founder and President of the Rencontres de Moriond,
de Blois and du Vietnam*

On the 80th anniversary of our friend K.K. Phua, I am very honoured to be invited by Professor Ngee-pong Chang, to write a tribute "as a pioneer in organizing the Rencontres de Moriond since 1966", to the long lasting efforts by KK in promoting research and new ideas in Asia and around the world.

I would like to express my great admiration and my deep appreciation for his extraordinary success in establishing the renowned World Scientific Publishing Company and his devotion to promoting science and education in Singapore, Asia and the world.

A thought that characterizes the action of KK and Doreen.

*From fragile seeds, their talent and vision have brought them into
century old trees that spread the fruits of knowledge in the world.*

Kim and I were lucky to meet KK and Doreen (they are one, inseparable!) and collaborate with them since the very first years of their successful odyssey and we are happy to share the memories that have marked our paths and our deep friendship.

It started on March 1977 when I invited K.K. Phua, a young and brilliant professor at Nanyang University to participate in the "Rencontres de Moriond". We shared our views and our wishes to develop particle physics

in the South East Asia particularly in Singapore and Vietnam and we started
to work together. As the International Conference of High Energy Physics
(ICHEP) was due to be organized in Tokyo in 1978, I proposed to KK to organize together the first High Energy Physics conference in Singapore with
the support and in the same spirit as the "Rencontres de Moriond" already
quite well known in the Particle Physics community. We invited our friends
to attend as most of them were on their way to attend the HEP conference
in Tokyo. Many prestigious particle physicists, mainly friends of Moriond
from Europe, United States, participated with enthusiasm during our meeting, among them Leon Van Hove, then General Directeur of CERN. After
the meeting, I asked KK to take care of the publication and distribution of
the proceedings which was done perfectly. Back in France, I kept in touch
with KK and the scientists of Nanyang and NUS Universities and invited KK
to University of Orsay as a visiting professor. During his stay, with Doreen
and the two children of about the same age as ours, our relations deepened
both through scientific exchanges as well as through friendship and family
exchanges. One day, inviting KK to visit my basement where we had an
office of a very small publishing company "Editions Frontieres" which took
care of the Moriond proceedings, I asked him if he could help us to print
our Moriond proceedings in Singapore which he accepted with enthusiasm.
We were far from thinking that these moments of meetings in a family setting were a trigger for the unthinkable adventure of KK and Doreen which
finally led to the birth of the famous World Scientific Publishing Company.

I am also very impressed by his devotion and his strong conviction and
energy to develop Fundamental Science in the South East Asia area.

I remember that just after the First High Energy Physics conference in
Singapore, he had already started the project of creating the South East Asia
Theoretical Physics Association.

Year after year, with the dazzling success of World Scientific Publishing company, KK created the Institute of Advance Studies which was the
host of numerous important scientific events: international conferences and
workshops, special anniversary events to express our recognition and appreciation of the elderly leaders of their respective fields: Chen Ning Yang,
Abdus Salam, Murray Gell-Mann . . .

With KK, we had always shared the idea of transmission of knowledge between generations as in the Rencontres de Moriond, with prominent and well-known scientists, among them Nobel laureates, who share with the young generations their experience and their passion in Science through formal and more importantly informal discussions in a relaxed atmosphere.

In this spirit, I am particularly happy to see the important success of the series of International Science Youth Forum organized by the Institute of Advanced Studies which allows high school and university students to exchange during five days with prominent Nobel laureate scientists, icons of the young generations. One of the most important features of the KK vision is the International cooperation with most important scientific institutions in the world.

With Vietnam, KK and Doreen have forged a very close relationship and are always there for the most important times. KK and Doreen were at the very first Rencontres du Vietnam in 1993 in Hanoi when Vietnam was still facing a difficult economic situation. They returned in 2013 to celebrate the 20th anniversary of the Rencontres du Vietnam and the inauguration of the International Centre of Interdisciplinary Science and Education (ICISE) that we created in Quy Nhon, following the spirit of Cargese, Erice, Les Houches, Aspen . . . and contributed strongly to the exchanges and collaborations between our two countries. Their presence at all these events strongly testifies to their faithful friendship.

The visit to Vietnam by an IAS delegation led by KK is one that I had the honour to host with the hope of creating a long term collaboration between scientific institutions in Vietnam and in Singapore especially with the Institute of Advanced Studies at Nanyang Technological University. The meeting with Professor Huynh Tan Dat, vice President of the National University in Ho Chi Minh city and presently Minister of Science and Technology of Vietnam and those in Hanoi with Professor Nguyen van Hieu, the most respected physicist in Vietnam, paved the way for cooperation between Singapore and Vietnam, permitting hundreds of Vietnamese scientists to visit and participate in summer schools, workshops and conferences organized in IAS and NTU. It is an important contribution to the development of

Science in Vietnam. I strongly hope this collaboration can continue in the future.

To conclude, from almost nothing, a small high energy physics conference, a cooperation with Editions Frontieres, a small proceedings publishing company, KK and Doreen have created the giant WSPC, and wholeheartedly devoted their lives to serve our community in developing science and education. I am very proud to have been with KK and Doreen in the very first part of this adventure and wish to express to both of them my sincere admiration and appreciation on this special event.

Warm wishes from Kim and myself to Doreen, KK and the family for good health, happiness and success forever in their achievements towards community development.

Tribute to Professor K.K. Phua

生
日
快
乐

Henry Tye

Hong Kong University of Science and Technology, Hong Kong

I first met KK on a memorable trip to China in the winter of 1979–1980. There were about 30 theoretical physicists (and some family members) on this trip, led by Profs. T. D. Lee and C. N. Yang. The weeks-long trip allowed us to get to know each other.

Later I learned that KK had started a publishing company in Singapore, namely World Scientific, focusing on physics publishing, which then grew to include a wider range in science and technology. I am honored to have served on the editorial board of the *International Journal of Modern Physics* and *Modern Physics Letters* for many years.

Through journals, conference reports, lectures and textbooks, World Scientific has provided a rich resource for readers of science and technology to grow and thrive in Asia and worldwide. I am amazed and truly impressed by KK and his wife Doreen in their unrelenting efforts in fulfilling their vision. The rapid rise of science and technology in Asia has benefited in no small part from the establishment of World Scientific Publishing Company.

Over the years, KK had also organized many international symposia and workshops at the Institute for Advanced Studies at Nanyang Technological University, bringing together the worldwide community of researchers for stimulating interactions to further the progress of science. Personally, I found those gatherings to be very fruitful.

KK's legacy will be remembered.

A Brief History of an Indian Initiative: ICTS–TIFR

生
日
快
乐

Spenta R. Wadia

International Centre for Theoretical Sciences,
Tata Institute of Fundamental Research,
Shivakote, Bengaluru-560089, India

Kok Khoo Phua and World Scientific

In 1986, a high-energy physicist Kok Khoo Phua (affectionately called KK), and his wife Doreen Liu started the publishing company World Scientific. I was very happy that an Asian publishing house of the type envisioned by them had arrived to break the monopoly over scientific publishing held by companies that were making products unaffordable in many institutions of higher learning in the developing world.

Over the years, World Scientific has grown into a major publishing house worldwide of scientific journals, books, and other educational publications. I was very happy to be associated with this enterprise and served on the editorial board of the *International Journal of Modern Physics A* and *Modern Physics Letters A* from inception. In the early years, I made many memorable visits to Singapore to discuss with KK and his staff and I was happy that they valued my suggestions and thoughts. On my part, I started publishing in World Scientific journals, and encouraging others to do the same.

I had arrived at the Tata Institute in Mumbai in 1982 from the University of Chicago. Among the many activities I joined in to improve our research and education, was organizing regular conferences and meetings. Those were the days when conference proceedings were popular, and KK was always generous about extending a helping hand. World Scientific published the proceedings of several meetings organized by the Tata Institute. The first one was the proceedings of the 1986 Winter School on "Strings, Lattice Gauge Theory and High Energy Phenomenology", which also contains an original unpublished paper by Yoichiro Nambu, "Duality and Hydrodynamics". These notes were prepared for an undelivered lecture at the Copenhagen High Energy Symposium, August 1970, and contain a detailed exposition of his ideas on string theory.

I was also happy to be invited to edit a World Scientific volume on "Large N limit in Quantum Field Theory and Statistical Mechanics" with Edouard Brezin in 1991, and in 2005 a book of articles by leading scientists commemorating the three revolutionary 1905 papers of Albert Einstein. World Scientific also published in 1999, "A Quest for Symmetry", which contains the selected works of Bunji Sakita edited by Keiji Kikkawa, Miguel Virasoro, and myself.

As a person, I always found KK to be very friendly and encouraging... bouncing off new ideas, impatient and in a kind of a hurry to reach his goals, only some of which I knew. He was also involved in organizing many scientific meetings in Singapore. My most memorable one was the conference to commemorate the 80th birthday of Murray Gell-Mann, his student Kenneth G. Wilson was also present and gave an amazing talk!

Time flies and KK is already 80! I would like to wish him many more wonderful and creative years.

In the following, I will describe another Asian initiative, the International Centre for Theoretical Sciences (ICTS) of the Tata Institute of Fundamental Research (TIFR) in Bangalore, India, with which I am deeply involved.

International Centre for Theoretical Sciences–Tata Institute of Fundamental Research: A Brief History

The idea of ICTS

By the late 1990s, the subject of string theory was doing well in India with some influential contributions and a fair-sized community. It seemed like a good opportunity to organize the annual strings meeting in India. There was good support towards this from the string theory community and we planned the meeting at TIFR in January of 2001. The initial funding was offered by Miguel Virasoro, then Director of ICTP Trieste, who remarked: "India deserves this". David Gross suggested that we "get Stephen" and that gave enormous publicity to string theory and fundamental physics in India. *Strings 2001* was indeed a watershed moment for the subject in India.

After this meeting in Mumbai, Edward Witten was invited to deliver a lecture at the Indian Academy of Sciences in Bangalore. During the visit to Bangalore, he wanted to visit the campus of Infosys Technologies in Bangalore, "a temple of modern India", rather than my suggestion to visit the Hoysala temples of Belur and Halebidu! The Infosys Campus was modern, and state of the art and it was a revelation to me that something like this was possible in India.

The idea of the ICTS was born in early 2001, after the successful organization of the *Strings 2001* conference at TIFR Mumbai and a visit to the campus of Infosys Technologies in Bangalore. The former boosted our confidence based on our achievement in fundamental physics and the latter assured us that institutional infrastructure and management of the highest international quality is possible in India. The combination of the highest quality science within a modern state-of-the-art campus, managed along modern management lines, inspired the basic idea of the Centre. However, ICTS was not to be a string theory institute but one where all of "science is one story". The partitioning of science into different areas is historical and perhaps a limitation of the human mind to process, comprehend and make sense out of so much disparate data. By the last century, it was clear that physics, chemistry and biology form a hierarchy of complexity in the sense meant by Herbert Simon.

In February 2001, I drafted a proposal to Infosys Technologies for the creation of a "Center of Excellence in Theoretical Sciences". It began with a critique of some aspects of the Indian science system and spelled out a vision for the Centre, its faculty, visitor programs, students, location, management, and funding. I did not hear back from them! The idea also did not find much traction at TIFR.

In 2004, discussions with David Gross in Santa Barbara gave a concrete road map and impetus to the creation of the Centre. We usually sat on a bench outside his office as cigars were not allowed in the building! David emphasized that Government funding is crucial and should be the main source with private funding filling in some crucial gaps. We discussed the people in India who should be brought on board and be convinced about the Centre. Following up on this meeting it took about two years before I was able to make a presentation to the Governing Council of the Tata Institute in 2006 and with the strong support of C. N. R. Rao and K. Kasturirangan, ICTS was approved in August 2007. I was given the responsibility of the first Director and asked to find resources, search for a suitable place to locate the ICTS campus, begin the international programs, build up the faculty and the administration. It felt like being thrown into the water without knowing how to swim! My colleague Avinash Dhar became convinced that ICTS was an important step forward for Indian science and joined in the effort. Sometime later we were joined by Uma Mahadevan Dasgupta, of the Indian Administrative Service on special duty at the Tata Institute, and by Mukesh Dodain who led the nascent administration at ICTS.

The ICTS began functioning from the Tata Institute in Mumbai holding its programs in institutions which were willing to host them. Some of the early programs were held at the International Centre in Dona Paola in Goa, IUCAA-Pune, JNCASR-Bangalore and the Indian Institute of Science-Bangalore.

The governance structure of ICTS consists of a Management Board, a Program Committee and an International Advisory Board of distinguished experts to guide the ICTS on all aspects of its growth. Their advice and encouragement are invaluable. The first Board members consisted of Michael Atiyah, Manjul Bhargava, Roger Blandford, Edouard Brezin, Michael Green, David Gross (Chair), M. S. Narasimhan, T. V. Ramakrishnan, Subir Sachdev, Ashoke Sen, Katepalli Sreenivasan, Raman Sundaram and S. R. S. Varadhan.

It took about two years and considerable effort to find a suitable city in India to locate the ICTS campus. Bangalore became a preferred choice because it had the academic ecosystem to sustain ICTS given the presence of the Indian Institute of Science (IISc), the National Centre for Biological Sciences (NCBS) and the Centre for Applicable Mathematics (CAM) of TIFR, among others, and the fact that the State of Karnataka offered a piece of land in North Bangalore. Once the location was fixed, the Centre and its budget were approved by the Atomic Energy Commission that was Chaired by Anil Kakodkar.

ICTS missions: Research, Programs, and Science Outreach

ICTS has a three-pronged mission. One mission is of course outstanding research and post-graduate education that is broadly organized into three main directions: complex systems that include statistical and condensed matter physics, fluid dynamics and turbulence, and physical biology; space-time physics that includes gravitational wave astronomy, quantum gravity and string theory, and related areas; and mathematics that includes geometry, PDEs, probability theory, data assimilation, and computing.

The other mission of the ICTS are its "Programs" of various durations in time that bring together physicists, astronomers, cosmologists, mathematicians, biologists, students and researchers from all over the world, under one roof, for various lengths of time to work together to solve the most challenging questions posed by nature, to discover the underlying structures across the sciences and to strive for the unity of knowledge. These programs include discussion meetings and named lecture series in honor of S. Chandrasekhar (Physical Sciences), S. Ramanujan (Mathematical Sciences) and A. Turing (Engineering and Biological Sciences). Special lectures like the Abdus Salam Memorial Lectures and the ICTS Distinguished Lectures are delivered by distinguished scientists as part of Science Outreach. ICTS programs also enable a platform for new, important emerging areas of science, mathematics and engineering.

Using the resources of these two main missions, ICTS anchors a vibrant program of Science Outreach that interfaces with civic society. This includes public lectures by eminent visiting scientists. It is important to share our work with civic society: with students so they can get interested in science; with people who are not scientists to have them participate in this incredible journey of exploration of the natural world including ourselves. Outreach is important as we seek the support of civic society for curiosity-driven research.

The inaugural Public Lecture was by Juan Maldacena on "Black Holes and the Structure of Spacetime" in July 2008. Other public events included lectures by Terence Tao on "Structure and Randomness in the Prime Numbers", Klaus Von Klitzing, "How Long is 1 Meter", William Philips, "Time Einstein and the Coolest Stuff in the Universe" and Lyman Page, "Observing the Birth of the Universe". "The Universe Unravelled: Cosmology, Gravitation and Elementary Particles" was a public event with Kip Thorne, Richard Bond, James Peebles and John Ellis.

Today ICTS organizes several science outreach activities. Kaapi and Kuriosity is a monthly Sunday lecture on an interesting science topic at the Nehru Planetarium in Bangalore; Math Circles is a national initiative at ICTS that identifies talented students and exposes them to the joy of doing mathematics at an early age; Einstein Lectures in partnership with institutions and colleges are delivered anywhere in India by distinguished scientists; Vishveshwara Lectures are delivered by leading scientists who are good communicators. The inaugural Lecture was delivered by Kip Thorne.

Moving from TIFR-Mumbai to the Bangalore campus

In December 2009, ICTS organized its Inauguration Conference, "Science without Boundaries" and the "Foundation Stone Ceremony" at the Indian Institute of Science in Bangalore. This conference informed the Indian and international scientific community about the ICTS missions and its commitment to a modern and transparent governance structure. The idea of the ICTS was succinctly articulated by Michael Atiyah in his Foundation Stone remarks:

> "... Science has the noble aim of trying to understand the natural world in human terms: to make sense of what we see. This brief phrase encapsulates both theory and experiment. What we see, in the broad sense, covers experiment and making sense is the task of theory. As the great French mathematician Henri Poincare said, science is no more a

collection of facts than a house is a collection of bricks: it requires theory to hold it together.

Theory needs a framework in which to develop and, as a mathematician, I believe that mathematics provide that unifying framework. As Galileo said the book of nature is written in the language of mathematics. Galileo was thinking primarily of mechanics and astronomy but, increasingly since his time, mathematics has provided the essential underpinning of ever-widening branches of science. As soon as a science moves from the qualitative to the quantitative, mathematics becomes indispensable.

Not only does mathematics provide the technical tools that all sciences require but, by its very nature, it acts as a unifying principle, integrating the diverse aspects of nature into an organic whole.

I am sure that mathematics, in all its various aspects, will play an important part in the future activities of this Center. In the complex modern world with the enormous challenges that we face, from climate change to energy, from poverty to water shortages, science provides the bedrock on which we can build our future. I am sure that this Center will play its part in guiding both India and the wider world in the years ahead."

ICTS Foundation Stone Laying Ceremony, 2009

In 2010, ICTS moved to its temporary campus in IISc from where it planned and built its new campus, continued its programs and academic activities, began recruiting its first set of faculties, and welcomed its first students and postdocs. Several institutions in North Bangalore including IISc, NCBS, and CAM lent a helping hand in the various tasks faced by ICTS.

We consulted colleagues worldwide, studied many architectural designs of institutions in India and abroad including the Newton Institute in Cambridge, and worked closely with the architects (Venkataraman and Associates) to realize the Centre we envisaged, keeping in mind the famous saying, "We shape our buildings, thereafter they shape us". The architecture, open spaces, and gardens contribute to an interactive academic atmosphere and a pleasant and comfortable stay on campus.

ICTS under construction, April 2013

As the campus was being built ICTS organized most activities in Bangalore. Some early activities include a multi-disciplinary meeting on "Scientific Discovery through Intensive Data Exploration", the "ICTP-ICTS joint School on Quantitative Biology", "Random Matrix Theory and Applications", "Winter School on Experimental Gravitational Wave Physics", "Asian Winter School on Strings, Particles and Cosmology", "Mathematical Perspectives on Clouds, Climate, and Tropical Meteorology", Ramanujan Lectures by Andrej Majda, a major math outreach activity, "Mathematics of Planet Earth", which included an exhibition on illustrating mathematics via simple hands-on models and a celebration of the Turing Centenary with a talk by Sydney Brenner, "The Architecture of Biological Complexity" and a discussion meeting, "On the Role of Theory in Biology".

We moved into the campus on June 2015 even while it was being completed and the landscaping was being done. The Campus Inauguration event was a conference titled "Science at ICTS" on 20 June 2015, which also included a public lecture by Manjul Bhargava, "Poetry, Drumming, and Mathematics".

By 2015 ICTS faculty had grown to consist of fifteen members, and the graduate school and postdoctoral programs were in place. *String 2015* was organized at ICTS in the Indian Institute of Science. The morning session on June 26th was held at the new ICTS campus and the afternoon session celebrated "100 Years of General Relativity".

The lecture halls and auditorium have state-of-the-art audio-visual and recording facilities for ICTS events which are then made available on YouTube. ICTS has a data centre (Turing Hall) that will eventually go up to 20,000 cores; it has an iconic library with an excellent collection; a guest house and residential facilities; a sports and wellness centre; and a computer centre and an efficient administration serves the Centre.

ICTS Library   ICTS Data Centre

The main financial support for the Centre comes from the Govt of India as part of funding for the Tata Institute of Fundamental Research. Besides this, ICTS engages with private foundations and philanthropy. It has been able to secure grants from the Airbus Corporate Foundation, the Simons Foundation, and the Infosys Foundation. There is an active resource development office that works with donors and "Friends of the ICTS" worldwide. ICTS also brings out a regular newsletter that is widely circulated.

In July 2015, I stepped down as Director and handed over the baton to Rajesh Gopakumar. Under him, the ICTS continues to grow and flourish in all its missions.

Activities at ICTS

The quantum of Programs, Discussion Meetings, Named Lectures, and Science Outreach has grown many fold in the new campus. ICTS presently has twenty-three regular faculty members. It has added researchers working in probability and statistics, and theoretical computer science to its Mathematics unit.

In keeping with our vision that "science is one story" there are no subject departments, but units that can interact with each other. ICTS faculty are domain experts in what they do, but their intellectual space is broader than their chosen field. This enables them to come together to solve problems that need diverse expertise. An example is the project to understand the mathematical foundations of the Indian monsoon — a weather system, which is very complicated and little understood, given India's proximity to the equator and and its special geographical characteristics of being a peninsula with high mountains in the north.

The research quality and keen involvement of the ICTS faculty forms the bedrock for its various activities like schools, conferences, workshops, discussion meetings and the science outreach activities described earlier. Some of the best people in the world, across the landscape of science, social sciences and humanities, visit and spend varying lengths of time at the ICTS.

ICTS carries out joint activities with various other institutions worldwide. ICTS has held, in conjunction with ICTP-Trieste, ten instalments of a successful *Winter School in Quantitative Biology* alternating between Trieste and Bengaluru. ICTS is the Indian node of the very successful *Kavli Asian Winter School on Strings, Particles and Cosmology*, which rotates between China, Japan, Korea and India. ICTS also held a joint workshop with IMPA-Brazil and ICTP on *Smooth and Homogeneous Dynamics*.

ICTS at Ten, January 2018

In January 2018, ICTS celebrated ten years of its existence by holding a major event "The ICTS at Ten". This was an opportunity to reflect on our journey since 2007, and to go ahead with renewed energy into the second decade. The theme of the meeting was "Celebrating the Unity of Science". Over the course of this meeting, speakers from around the world gave talks with broad perspective across several different themes in the theoretical sciences: Astrophysics and Cosmology, String Theory and Quantum Gravity, Mathematics, Theoretical Computer Science, Condensed Matter and Statistical Physics, Physical Biology. These areas reflect ICTS' profile in 2018 as well as the directions it would like to grow into in the coming years.

The same year, ICTS had a comprehensive review undertaken by a distinguished international committee consisting of Beverly Berger, Curtis Callan (Chair), Bertrand Halperin, Kavita Jain, Itamar Procaccia, and Ashoke Sen. The committee concluded that *"ICTS is a powerful new asset for Indian science and science education. Much remains to be achieved, but a solid foundation for a future of great importance to Indian science and Indian society has been laid. ICTS has a unique multi-faceted mandate: to perform interdisciplinary research in theoretical science at the highest level across a broad intellectual frontier; to educate a new generation of PhD scientists in these disciplines; to facilitate the diffusion of frontier scientific developments and theoretical techniques across the Indian scientific enterprise through focused topical programs; to excite and inspire the public, especially the young who will be the scientists and technologists of India's future, through outreach programs. It is our assessment that ICTS, despite its youth as an institution, is doing an impressive job on all four fronts".*

In May 2018, ICTS initiated the D. D. Kosambi Lectures devoted to the social sciences, the arts, and the humanities in honor of the first Professor of Mathematics at Tata Institute, who made pioneering contributions to the methods and study of ancient Indian history. The inaugural lecture was delivered by Pratap Bhanu Mehta. In February 2020, ICTS initiated the Madhava lectures devoted to the history of mathematics, science, and technology, in honor of the great Indian mathematician of the Kerala School that invented rudiments of the Calculus. The inaugural lectures were delivered by P. P. Divakaran.

ICTS is involved in the Bengaluru Science and Technology (BeST) Cluster Initiative of the Govt. of India which is a collaborative ecosystem in which scientists, engineers, social scientists, and entrepreneurs working in academia, government labs, and industry, identify and collaborate towards solutions for some socially relevant problems. BeST has set up teams to be engaged in areas related to health, agriculture, urban transportation, climate, quantum science, materials, and jet engines.

During the Covid pandemic, the campus was closed for almost two years. ICTS activities shifted online. During these hard times ICTS created Covid Resources which has commentary on the COVID-19 data analyzed by ICTS scientists, as well as links to useful online material.

ICTS Campus now

The Future

Since its inception, ICTS has achieved some measure of success in its three missions! Its programs have had a significant impact on Indian science; it is an international hub of science; its faculty (presently twenty-three) has made widely recognized contributions and its science outreach has become a fixture for science enthusiasts in India.

ICTS is considered by many as one of the best institutions of its kind in the world. A distinguished visitor to ICTS from Caltech said his next destination was the ICTS of the North (IAS Princeton).

Going forward, a successful institution like ICTS should grow in size and expand the scope of its activities to serve a larger number of researchers and students. In this way it will become a greater national asset for India. Its present areas of research could be augmented to include, exploring areas of evolution, learning and cognition, artificial intelligence and computational mathematics to name a few. In another vein, progress in quantum science and mathematics will enable deeper insights into the workings of the quantum world and provide a paradigm shift in computing, sensitive devices and engineering, which can feed back and enable us to make fundamental discoveries about the nature of the universe and life.

Further, into the future ICTS could include experimental science, engineering and social sciences and humanities, and strive to be a research university that integrates knowledge and human values. India is a country of over a billion people trying to emerge from its inequitable past. It needs many, many such institutions of higher education and research!

A Great Science Promoter

生
日
快
乐

Yifang Wang

Institute of High Energy Physics, P. R. China

It is hard to believe that our dear friend KK is turning 80 years old. I still remember when I first met him in IAS-NTU for a workshop, when he was in his 60s. He seems to have not changed since then, always energetic at any given moment, full of passion, completely devoted to science, its publication, and international collaboration.

I heard the name of Kok Khoo Phua and his publishing house when I was a student, long before I met him. In 1990, the *International Conference of High Energy Physics* was held in Singapore. It is the second time of this conference series in Asia, just after Japan and the first and the only time in South East Asia. I had my thesis work being part of the presentations by Prof. Samuel Ting on behalf of the L3 experiment at the Large Electron Collider (LEP). I was eager to go, but too young for it to be possible. This conference initiated and organized by KK was a great success and won very high praise from many attendees. I got an invitation after I became senior enough, I was very excited and immediately decided to go. Indeed, it was a great experience which triggered my visits to IAS for all possible occasions in the next 20 years. During the workshop and conferences at IAS, I met many old and new friends, learnt a lot from talks and discussions, and such memories will remain in my heart forever. Indeed, IAS at NTU has become a model for many universities, including the one at the Hong Kong University of Science and Technology, which I also frequently visit.

On several occasions, hot debates lead to interesting memories with or without consensus. A famous example is the discussion on several occasions about the future of high energy physics in China and in the world, in particular whether China should build a Higgs factory, called the Circular Electron-Positron Collider (CEPC). KK organized through World Scientific to publish several books on it, both in English and in Chinese. In 2020, the European high energy physics community with international inputs concluded their strategy to put the Higgs factory as the top priority. Scientific conclusion is now clearer even if the open question still remains: who can and will build it? The Covid-19 pandemic unfortunately prevented KK and IAS to organize further discussions in a face-to-face mode, otherwise better understandings and more collaborations may have occurred between Asia and Europe.

Everybody knows that World Scientific is a great publishing house which helped to promote science and science communications in an unprecedented way. The establishment and the operation of World Scientific is of legend which will be told by many others, I guess. I have had my personal experiences to publish proceedings and books. Our sentiments are always very positive, and our feelings are always smooth and comfortable. In particular, at the beginning of the upgrade of the Beijing Electron-Positron Collider (BEPCII), we planned to publish a handbook to describe physics goals for students, ourselves and funding agencies. A few years of efforts by about a hundred authors concluded the book — *Physics at BESIII*, while its publication has become an issue. We would like to have sufficient influence internationally, but at an affordable cost. This was 15 years ago and our budget was far less than today. Among all publishers, World Scientific gave us a good and the best offer. In the end, this was a very successful publication with more than a thousand copies printed and hundreds of students use it for their study of tau-Charm physics at BEPCII. We really appreciate the help by KK and World Scientific.

KK is a key player in the region for the international cooperation on science, in particular on physics. He founded or co-founded a number of science organizations such as Association of Asia-Pacific Physical Societies

(AAPPS), South East Asia Theoretical Physics Association (SEATPA), and published their newsletters. I was once invited by KK to attend the South East Asia physical society meeting, and met many new friends. It helped us to build collaborations in particle physics with countries like Thailand, Malaysia, Vietnam, etc. and our JUNO neutrino experiment at Jiangmen, Guangdong province in south China now have many members from South East Asian countries.

KK established IAS at NTU which has now become a model, at least in Asia. KK established a publishing house which changed drastically the mode of publication in science. Scientists, especially those from developing countries, benefited a lot from it. KK established a number of associations in Asia which boosted science in the region. KK is truly a promoter of science and science communications. We all wish for KK to continue the efforts, never retire and have a long and healthy life.

KK and His Scientific Contributions in Asia

Andrew Wee

Department of Physics,
National University of Singapore

Introduction

Science has been defined as both a process and a product. It involves "careful observation, applying rigorous skepticism about what is observed, given that cognitive assumptions can distort how one interprets the observation" [1]. It involves formulating hypotheses, via induction, based on these observations; experimental and measurement-based testing of deductions drawn from the hypotheses; and refinement of the hypotheses based on the experimental findings.

> *"The scientific process is a rigorous method of investigating a specific fact or event, while the products of science comprise a body of knowledge . . ."* [2]

KK, as Kok Khoo is fondly referred to, is well known to all as a high energy physicist who has dedicated his lifelong career to physics publishing. KK has been contributing to the communication of physics, benefitting the global physics community as reflected in the name of his publishing company "World Scientific". To those of us who know him well, we know of his specific interest in enriching scientific progress in Asia, and Singapore, in particular.

Journal publications

As a surface and nanoscale scientist myself, I shall share my personal anecdotes on how I became involved in the process of science communication and journals in the field of nanoscale science. In the year 2002, KK had the foresight of the significance of the fledgling field of nanoscience, driven by the inventions of nanoscale imaging techniques such as scanning probe microscopies. KK approached me to be the inaugural Managing Editor of the *International Journal of Nanoscience*, and so I did for several years (Fig. 1). *International Journal of Nanoscience* is an interdisciplinary peer-reviewed scientific journal that covers research in nanometer scale science and technology, with articles ranging from the "basic science of nanoscale physics and chemistry to applications in nanodevices, quantum engineering and quantum computing" [3]. Since then, I have also been put on the Editorial Board of *Surface Review and Letters* (Fig. 2). The scope of this journal covers "a broad range of topics in experimental and theoretical studies of surfaces and interfaces" [4].

The *Singapore National Academy of Science (SNAS)* is Singapore's scientific learned society, and has long aspired to have its own journal. Finally, SNAS launched its own journal as *COSMOS*, the name reflecting its global prominence. Some years later, I suggested that it be renamed the *Proceedings of SNAS* to better reflect its content and scope (Fig. 3). The *Proceedings of the Singapore National Academy of Science* "publishes invited review and research articles with the aim of promoting interdisciplinary research in Science and Mathematics. Each volume, published twice a year, focuses on a specific topic or field with a Singapore bias, accessible to researchers from other scientific disciplines" [5].

For KK's physics discipline, World Scientific and the Institute of Physics Singapore jointly launched *The Physics Educator* in 2019 (Fig. 4). KK has long valued physics education from preuniversity onwards, and so the focus of the journal is the "teaching and learning of physics and related topics at the secondary school, high school, junior college and the introductory undergraduate level" [6]. The aim of the journal is to enhance the teaching of physics in the classrooms through new approaches, insights and

Fig. 1. *International Journal of Nanoscience.*

Fig. 2. *Surface Review Letters.*

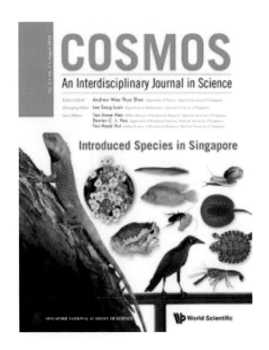

Fig. 3.   *COSMOS / Proceedings of the Singapore National Academy of Science.*

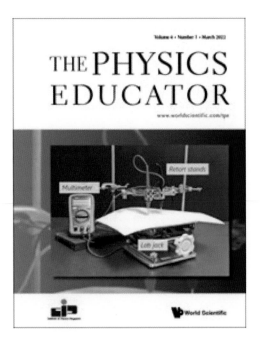

Fig. 4.   *The Physics Educator.*

demonstrations, and to provide a forum for discussions on educational and curriculum developments. This constitutes making a direct contribution to physics education in the classroom, especially within an Asian context.

Books

KK is also passionate about publishing books written by famous scientists. World Scientific is most renown for its series by Nobel laureates [7]. In 2021, I have had the privilege to coedit a book with Konstatin Novoselov, currently employed by the National University of Singapore. The Nobel Prize in Physics 2010 was awarded jointly to Andre Geim and Konstantin Novoselov "for groundbreaking experiments regarding the two-dimensional material graphene" [8]. Besides contributing two chapters, Novoselov also designed this eye-catching graphene-inspired cover image (Fig. 5). This book focuses on the molecular interactions on two-dimensional materials, which extend the understanding of surface physics and chemistry at the graphene interface.

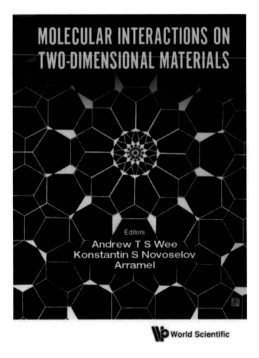

Fig. 5.   Edited book on *Molecular Interactions with 2D Materials*.

Conclusion

KK has indeed been an inspiration for generations past and present with his efforts to promote science in Asia with both journal and book publications. We wish him well on his 80th birthday, and hope that he will inspire many more generations to come in Asia through these publications.

References

[1] https://en.wikipedia.org/wiki/Scientific_method
[2] *J Nurs Res.* 2019 Feb; 27(1): 1.
     Published online 2019 Feb 5. doi: 10.1097/JNR.0000000000000315
[3] http://www.worldscinet.com/ijn/mkt/aims_scope.shtml Journal Aims and Scope
[4] https://www.worldscientific.com/worldscinet/srl
[5] https://www.worldscientific.com/page/psnas/aims-scope
[6] https://www.worldscientific.com/page/tpe/aims-scope
[7] https://www.worldscientific.com/page/nobel
[8] https://www.nobelprize.org/prizes/physics/2010/summary/

World Scientific in my Academic Career — Celebrating the 80th Birthday of K.K. Phua

生
日
快
乐

Yue-Liang Wu

International Centre for Theoretical Physics Asia-Pacific (ICTP-AP, UNESCO)

World Scientific Publishing Company (WSPC) has just celebrated its 40th anniversary. The amazing achievements of WSPC are the best gifts to celebrate the 80th birthday of K.K. Phua who initiated WSPC in the 1980s. It is a great pleasure to take this opportunity to look back on my association with World Scientific and the friendship with K.K. Phua, as my academic career grew in the same period as WSPC developed rapidly.

Around the time when Prof. K.K. Phua established World Scientific, I started my graduate study under the supervision of Prof. K. C. Chou (Zhou Guang-Zhao). The year 1986 came to be a special year when my paper with Prof. Chou was published in the first volume of *Modern Physics Letters A* (MPLA), "Effective action of σ-model anomalies with external gauge field," *Mod. Phys. Lett. A* **1**, 23 (1986). It was an invited article from WSPC's founder Prof. Phua who had written to Prof. K. C. Chou to contribute and support the newly launched journal of WSPC. While preparing such an invited article, I learnt for the first time from Prof. Chou that K.K. Phua is also a theoretical physicist, but he had chosen an alternative way for giving back to the academic community, promoting scientific innovation and knowledge to society by establishing Asia's first scientific publishing company. In view of the success and rapid growth of WSPC, Prof. Chou said to me that a theoretical physicist should be capable to do well any meaningful thing that he wishes to do. Prof. Chou also mentioned to me several theoretical physicists who had made great contributions beyond the frontiers of science.

My first meeting with Prof. K.K. Phua was in 1990 when I participated in the *25th International Conference on High Energy Physics (ICHEP 1990)* held in Singapore, where I delivered an invited talk about "ε'/ε and top quark", the work was done along with Prof. E. A. Paschos when I was a postdoc in Dortmund University. Later on, I received a letter from Prof. Phua who invited us to write a brief review article for MPLA, "Correlations between ε'/ε and heavy top quark," *Mod. Phys. Lett. A* **6**, 93 (1991), in which we emphasized for the first time on the importance of electroweak penguin contribution to ε'/ε when top quark is heavy (after about four years, top quark was discovered by FermiLab to be a really heavy one).

One year later, I came across more progress in the studies on direct CP violation and heavy top quark, which was immediately published in *International Journal of Modern Physics A* (IJMPA), "A simultaneous analysis for $\Delta I = 1/2$ rule, direct CP violation, K0-K0bar and B0-B0bar mixing and an optimistic prediction for top quark mass m_t, CP phase δ and ε'/ε," *Int. J. Mod. Phys. A* **7**, 2863 (1992). It was such an interesting observation that I was invited again in 1992 to present a talk about "Recent theoretical development on direct CP violation ε'/ε" at the *Int. Conf. on High Energy Physics (ICHEP 1992)* held in Dallas, USA. In the talk, I addressed confidently the significant nonperturbative contributions in kaon decays, thus enabling us not only to explain the long-time puzzle on isospin $\Delta I = 1/2$ rule, but also bring a consistent prediction on direct CP Violation ε'/ε which was found to be related to the sensitivity of experiments at that time and should be measurable by improved and more precise experiments. The confirmation of such a theoretical prediction should enable us to test the CP-violating source in standard model and probe super weak theory as proposed by Lincoln Wolfenstein. For that, I was invited in 1993 by Prof. Wolfenstein to visit Carnegie-Mellon University and explore together CP-violating mechanism and new CP-violating sources (*PRL* **73** (1994) 1762 and *PRL* **73** (1994) 2809). After about ten years, two improved experiments at CERN and FermiLab truly led to the direct CP violation ε'/ε in kaon decays.

Besides publishing articles in the journals by WSPC, I was also invited by Prof. K.K. Phua to publish Proceedings. The first Proceedings on *The Frontiers of Physics at Millennium* was produced by WSPC in 1999, which kept detailed records on the most excellent talks by outstanding Chinese physicists at the Symposium held at the Institute of Theoretical Physics, Chinese Academy of Sciences. Such a symposium was initiated after I returned to China with the support of most distinguished senior physicists in China, and recognized as the first Fall Meeting of Chinese Physics Society (1st CPS Fall Meeting). Two years later, I published the *Proceedings on Flavor Physics*, which recorded the excellent talks at the *First International Conference on Flavor Physics (ICFP 2001)*. Such a conference was so successful that it led to a series of international conferences first initiated in China.

So far, more than 30 of my publications have been printed with WSPC since my first paper appeared in *MPLA* in 1986. Being a theoretical physicist, I cannot but say that World Scientific has played a critical role in my academic career, in particular, publishing those papers on initial creative ideas from quanta to cosmos. Over the last year, I have had two special issues printed with WSPC, namely: "Taiji program in space for gravitational universe with the first run key technologies test in Taiji-1" (*Int. J. Mod. Phys. A* **36** (2021) 11, 12); "The foundation of the hyperunified field theory I – Fundamental building block and symmetry and The foundation of the hyperunified field theory II – Fundamental interaction and evolving universe" (*Int. J. Mod. Phys. A* **36** (2021) 28). I am delighted to take this opportunity to thank K.K. Phua for his support.

On the occasion of K.K. Phua's 80th birthday, I also greatly appreciate his support in publishing the book *Selected Papers of K. C. Chou* which was dedicated to the 80th birthday of K. C. Chou (Zhou Guang-Zhao) in 2009, it was his support that helped me to obtain very meaningful book prefaces from Prof. C. N. Yang and Prof. T. D. Lee.

It is my honor to share my own experiences about World Scientific with the entire academic community, and see how Prof. K.K. Phua has made great efforts to support young researchers and give back to the academic community through World Scientific. I would be pleased to dedicate this short article to celebrate the 80th birthday of K.K. Phua.

A Generous Giant — Professor K.K. Phua

生
日
快
乐

Nai-Chang Yeh

Department of Physics, California Institute of Technology,
Pasadena, California 91125, USA
ncyeh@caltech.edu

Professor Kok Khoo Phua, fondly known to his colleagues as KK, is a giant in the physics community who has been highly accomplished in the promotion of physics research and education worldwide. I had the good fortune and honor to co-chair the 8th Joint Meeting of Chinese Physicists Worldwide (OCPA8) with KK, and so had close cooperation and interactions with him and his team for more than a year on planning the large international event held in Singapore on June 23–27, 2014. The experience was an absolute delight because KK and his team were thoughtful, cooperative, and generous in lending their support. This article describes my fond memory of the OCPA8 Conference, which serves as my tribute to KK on his 80th birthday for his previous great service as the Co-Chair of OCPA8 and for his tireless and exemplifying contributions to the physics community overall.

1. Introduction

It seems that every physicist of Chinese heritage would naturally come to know Professor Kok Khoo Phua because he is the founder of the internationally renowned Institute of Advanced Studies (IAS) at Nanyang Technological University and the Chairman and Editor-in-Chief of World Scientific Publishing Company. I recalled that my first direction interaction with KK was when he invited me to write a review article in the area of high-temperature superconductivity, which I deemed as an honor and so was gladly obliged.[a] In 2013, when I was elected to the Presidency of the Overseas Chinese Physicists Organization (OCPA, which is now renamed

---

[a]N.-C. Yeh and A. D. Beyer, Unconventional low-energy excitations of cuprate superconductors, *Int. J. Mod. Phys. B* **23**, 4543–4577 (2009).

as the International Organization of Chinese Physicists and Astronomers and is still abbreviated as the OCPA),[b] the organization had decided to hold the 8th OCPA international conference in Singapore from June 23 to 27 in 2014, with Professor K.K. Phua taking the leading role in hosting the event in Singapore on behalf of IAS at the Nanyang Technological University and the National University of Singapore. Therefore, I had the opportunity to represent the OCPA to co-chair the 8th Joint Meeting of Chinese Physicists Worldwide (OCPA8) with Professor K.K. Phua and to cooperate with him and his team on the planning and running of OCPA8 from late 2012 to 2014.

2. Joint Planning for the OCPA8 Conference

The OCPA8 Conference in 2014 was one of a series of OCPA Conferences held in Shantou (1995), Taipei (1997), Hong Kong (2000), Shanghai (2004), Taipei (2006), Lanzhou (2009), Kaohsiung (2011) and Beijing (2017). The overarching objective of OCPA8 in Singapore was to highlight the latest scientific breakthroughs and achievements by physicists and astronomers worldwide, especially since OCPA7, to help promote physics research and teaching in Chinese societies and beyond, and to enhance international collaboration. Given the broad scope and the large number (many hundreds) of international participants of the OCPA Conferences, advanced and detailed planning by the scientific program committee and local organizing committee was essential. KK and his team in Singapore started working closely and tirelessly with me and the OCPA scientific program committee since November 2012. It was a true partnership every step of the way: we jointly designed and finalized all the scientific and educational programs, meeting formats, and the list of invited speakers; KK and his team secured generous financial support and arranged for all the local logistics for the conference, while OCPA joined forces with the American Physical Society (APS) to provide funds to support the poster session and poster awards for young researchers. The leadership demonstrated by KK was essential for the smooth planning and ultimate success of OCPA8.

---

[b] Homepage of OCPA: https://www.ocpaweb.org/home.

3. Highlights of the OCPA8 Conference

The theme of OCPA8 was "Looking forward to the Quantum Frontiers and Beyond", and the range of topics covered in the five-day scientific program included Accelerator Physics, Atomic & Molecular Optical Physics, Astronomy and Astrophysics, Biophysics, Chemical Physics, Computational Physics, Condensed Matter Physics, Gravitation and Cosmology, High Energy Physics, Interdisciplinary Physics, Nuclear Physics, Plasma Physics, Quantum Computing, Statistical and Nonlinear Physics, Science Education, and Women in Physics. Given the breadth of topics, the co-chairs and program committee members collectively decided to vary the format of OCPA8 slightly from those of the previous OCPA conferences so that all oral presentations were invited talks, and that more plenary presentations and fewer parallel talks than those in the past were scheduled. This format thus enabled the latest research highlights in different disciplines to be better conveyed to all attendees in the plenary session, and that the education, outreach and international components of the conference were able to engage a broad spectrum of the conference attendees in the setting of plenary talks and plenary roundtable discussions. To compensate for the reduced number of oral presentations, we placed strong emphasis on the poster presentations, and implemented for the first time the "OCPA-APS Outstanding Conference Poster Award", which was jointly sponsored by the American Physical Society (APS) and OCPA. Up to three outstanding contributions to the poster presentations in each subfield were selected by the poster award committee onsite and were honored at the conference banquet. Each award constituted a cash award and an official certificate, both jointly issued by the APS and OCPA.

All the invited speakers for both the plenary and parallel sessions delivered excellent talks on not only their own research but also the status of their perspective fields, which stimulated lively discussions and exchanges among the conference participants. In particular, Nobel Laureate Professor Chen-Ning Yang (Tsinghua University) kicked off the OCPA8 Conference with an inspiring plenary talk entitled "Conceptual origin of Maxwell's equations and field theory"; Nobel Laureate Professor Carlo Rubbia (CERN) in his plenary talk on the second day, entitled "Neutrinos: a golden field of astro-particle physics", gave an informative overview and further described various exciting ongoing developments in the field of neutrino physics; APS President Professor Michael Turner (University of Chicago) opened the third day plenary session with a fascinating perspective on "The big mysteries of cosmology", which was very well received by all conference attendees from different physics subfields.

In addition to the scientific program, another important highlight of OCPA8 was the plenary sessions of Science Education Roundtable, Women in Physics Roundtable, and International Collaboration Roundtable. We invited eminent scientists and higher-education leaders as the speakers and panelists for these sessions, which actively engaged the conference attendees in the discussions of these topics. Additionally, Professor Chen-Ning Yang graciously delivered a public lecture on the topic of "My experience as a student and researcher" on the third day to a group of enthusiastic local high school students and teachers, which was truly inspirational.

Besides the intellectual feast, the local organizing committee provided excellent treats for the welcome reception, coffee breaks, lunches, and conference banquet. There was also an evening pizza party, jointly sponsored by the APS and OCPA, for the 2.5-hour poster session. Additionally, 2-hour research lab tours in the afternoon of the third day and a 4-hour city tour in the afternoon of the fifth day were arranged by the local organizing committee. These social activities facilitated excellent opportunities for conference attendees to mingle and interact. The hospitality of IAS and NUS led by Professor K.K. Phua made the entire experience of the OCPA8 Conference truly memorable. Professor Phua even took a few of us for a special visit to the headquarters of the World Scientific Publishing Company so that we had first-hand observation of the operation of the impressive enterprise. The feedback that we received from all participants of the OCPA8 Conference was excellent. Professor Phua and his team together with all OCPA program committee members and the APS leadership certainly deserved all the credit for the success of OCPA8.

4. A Few Words of Tribute

It is hard to believe that almost 8 years have passed by since KK and I co-chaired the OCPA8 Conference, even though the fond memory is still vivid in my mind. Through our cooperation, I was able to appreciate the amazing energy, leadership, and conviction of KK to always "do the right thing". He has been highly accomplished in promoting physics research and education worldwide through the founding of a world-class IAS and all the reputable and influential publications from World Scientific Publishing Company. Moreover, his tireless effort has undoubtedly elevated the international stature of physicists of Chinese heritage. It has been a great pleasure to know and work with KK, who is a fierce and visionary leader, a gentle and generous giant, and a wonderful colleague. Happy 80th Birthday, KK! I wish you continuous success and thank you for all you have done for OCPA and the physics community worldwide.

To K.K. Phua at 80

生
日
快
乐

Kenneth Young

Department of Physics,
The Chinese University of Hong Kong, Hong Kong, China

It is a pleasure to offer the heartiest congratulations to my friend K.K. Phua on his 80th birthday and to recall his marvellous contributions to physics and science in general in Asia.

Time flies, and my memory is not reliable in details, though the broad brushes remain vivid. I recall first getting acquainted with KK when he paid a short visit to Hong Kong in the mid-1970s. We met through a mutual friend, and discovered many common interests. It was perhaps 45 years ago, and the world was different in ways that the younger generation would find difficult to appreciate. Of course, there were no internet and fax machines were not yet generally available; long-distance calls were expensive and inconvenient. For most scientists in Asia (with the possible exception of those in a few major centres in Japan), the sense was strong of being far removed from the mainstream of what was happening in science, even for relatively well-connected places such as Singapore and Hong Kong. The challenges were especially acute in high-energy physics, which was just undergoing a revolution with field theory becoming dominant, displacing more phenomenological approaches which had been common for a decade or more. One would await the arrival of academic journals by surface mail; collaborations, except in the same city, had to rely on handwritten messages and notes exchanged by post. In those circumstances, the opportunity to come into contact with another physicist in the region, interested in broadly similar fields, was received with pleasure. But more importantly, for KK it provided the motivation and the impetus to contribute, in very significant ways, in building bridges and institutions that would help to change the situation of regional isolation.

The next encounter that I recall was the 1978 *International Meeting on Frontiers of Physics*, held in Singapore. A regional meeting was a rarity in those days, and a much welcomed one, since travel further afield was not something quite so easily entertained. Although there were more senior physicists in Singapore, KK, then in his mid-30s, was evidently the driving force behind the event. As I said, my memory is faulty on details, but in this instance, I can fix the date well, since in my list of publications I do find an article in the Proceedings of that meeting, published under the auspices of the Singapore National Academy of Science, and labeled as "edited by K.K. Phua *et al.*". In retrospect, that must have been KK's first foray into the world of scientific publishing, and a harbinger of things to come.

We must have met a few times in the succeeding years, but the next entry I can find in my own records is another conference presentation, this time published in the *Proceedings of the First Asia-Pacific Physics Conference*, Singapore, 1983. This meeting was in many ways a successor to the one in 1978, but several features are worthy of note. First, there was a very conscious effort to reach out beyond Southeast Asia and to connect to the larger scientific communities in Asia. Thus, this volume was edited by Arima, who must have been the chairman of the meeting. Second, this meeting was auspiciously named the First Asia Pacific Physics Conference (APPC1), starting a series, with the latest one to be APPC15 in Korea, to be held in August 2022, unfortunately mainly online. This whole series, which has become a notable event that helps bring Asian physicists together, was in effect started by KK and driven by his enthusiasm and organizing abilities.

However, the third and possibly most significant feature of note for the event in 1983 is that the volume of Conference Proceedings is listed as published by World Scientific Publishing Co. (WSPC) in 1984. WSPC must have been just established, very much a fledgling organization.

The tradition developed, there in APPC1, that during the latter parts of each conference, representatives from the different physics communities would meet to decide the approximate date and the host for the next one. APPC2 was held in Bangalore in 1986, hosted by the Indian Physical Society, with Proceedings published by WSPC in 1987. I missed that meeting, but was much involved in organizing APPC3 in Hong Kong in 1988.

It is known (and for example, clearly displayed on the website of APPC15) that APPCs are organized by the Association of Asia-Pacific Physical Societies (AAPPS) together with a local host, usually a local physical society. But it is not widely known that APPCs predated AAPPS. The decision to form such a Society was taken during APPC3 in Hong Kong. C. N. Yang (participating through his role in Hong Kong) was the central figure, together with M. Konuma of Japan. They were to become the founding President and Vice-President of AAPPS, with Konuma succeeding Yang as the President for the next term. Both KK and I were active in supporting the process, and we were both on the Council for a number of years thereafter. Through those roles, and through attendance at a number of APPCs thereafter, I had many occasions to interact with KK, and to learn from him. WSPC, by then growing steadily, always allowed its resources and its network of connections to be leveraged.

The wisdom of KK can be found in some features of the Constitution of AAPPS, which bear pointing out even after these years. There was a conscious decision to be inclusive, to build a community that transcends issues that might divide different parts of the Asia-Pacific region. Thus, Members of AAPPS are Physical Societies (not individuals), but carefully phrased so that they are not necessarily national societies. Thus, for example, under AAPPS auspices, APPC7 was held in Beijing 1997, APPC8 was held in Taipei in 2000 and APPC11 was held in Shanghai in 2010. As a Chinese, I am of course pleased to be part of an organization that encourages this kind of dialogue, and KK, with strong affinities for his ethnic and ancestral origins, evidently also has similar feelings. The AAPPS Constitution also provides for Council members to be individuals, so that most decisions can be reached without having to seek formal endorsement from Member societies.

In the meantime, WSPC continued to grow steadily in size and influence, with Doreen most capably taking charge of the managerial and business side. The history of WSPC will be known to many, and will no doubt also be shared by other contributors to this *Festschrift*. I shall only recount one episode that I know about, which illustrates KK's compassion and support for like-minded individuals. Shang-Keng Ma of UCSD had spent two sabbatical years in Tsinghua University in Hsinchu, Taiwan, giving lectures on statistical mechanics which then evolved into a textbook written in Chinese. The book came to the subject with a fresh perspective, and was very educational even for those who supposedly already knew the subject. Unfortunately Ma was stricken by a serious illness, and eventually died in 1983, in his mid-40s. There was a proposal to render the book into English for a broader audience, and the project was taken up by WSPC. KK then sought my help to source a translator. That was when I came to know about the book, and of Ma. I was immediately struck by Ma's goal of bringing this sort of science, at once cutting edge and accessible to beginning graduate students, to the next generation — and doing so in their own language, in a fine balance between global excellence and indigenisation of science. I do not think that KK knew Ma, at least not well; neither did I. But I am sure both of us admired Ma's honourable sentiments. In the end, I was able to introduce M. K. Fung, an expert in statistical mechanics then at Tamkang University. Fung did a very nice translation and the book, in English, was published by WSPC. I took some interest in the project, and gained much insight into the subject (whereas I have to confess that I had hitherto only known how to calculate). This episode, no doubt one of many of the same sort, testifies to KK's enthusiasm to support projects and people that his excellent judgement regards as worthy.

There is yet another strand of KK's contributions that should not go unrecognized. For more than a decade until the last few years, KK has led the Institute of Advanced Studies (IAS) at Nanyang Technological University (NTU), even as he was running WSPC. To many of us in the region, we know of IAS through many of its conferences and workshops that brought high-level science to NTU, to Singapore and indeed to the entire region.

The emphasis on basic science was a welcome complement to the enviable developments in applied science and engineering, and a timely one for a region that now has the wherewithal to invest in the higher aspirations of the human spirit. For example, I recall having attended, at KK's invitation, events honoring various birthdays of C. N. Yang and Murray Gell-Mann, and one in memory of Feynman; for me, those events were all the more precious because I have the good fortune to know all three of these giants of physics. KK's interest also included the direction of higher education, and he arranged a conference in 2015 on Liberal Education in Asia, again held at IAS.

Even now, our paths are connected in two projects of interest. We are both (with four others) on the editorial committee for two volumes celebrating the 100th birthday of C. N. Yang, hopefully to come out in 2022. The first volume, consisting of essays of a personal nature and mostly in Chinese, will be published by Tsinghua University Press in Beijing; the second volume, consisting of scientific articles, will be published by WSPC. At the same time, WSPC has agreed to publish a set of four volumes about the first 15 years of the Shaw Prize, also scheduled for this year. It will be a miracle if either of these projects will make money for WSPC, and KK's enthusiasm in supporting these endeavours illustrates how he recognizes values larger than commercial success.

In a real sense, KK has accomplished several parallel careers in the last few decades. Now that the next generation is gradually taking over some of the responsibilities at WSPC, we all wish that KK will enter another phase of happy yet productive life, and contribute to science and society in yet other ways, as an elder statesman with stores of wisdom to share.

My 40+ Year Relationship with K.K. Phua and World Scientific

生
日
快
乐

A. Zee

University of California,
Santa Barbara, CA 93106, USA

On this doubly happy occasion of K.K. Phua's 80th birthday and World Scientific Publishing's 40+ birthday, I am pleased to write a few words for the commemorative volume. My memory of what transpired forty plus years ago is necessarily a bit vague, but I have a clear mental image of K.K. Phua telling me on a train in China about his plans to start a company publishing physics books and asking me to "help out". That was in either late 1980 or early 1981, because World Scientific Publishing, now a household name in physics circles but evidently unknown at that time, was registered on December 12, 1980 and started actual operation in early 1981 with a staff of five in a tiny office. They certainly worked fast, indeed at almost tachyonic speed compared to plodding pace of the ancient publishing houses in the US and in Europe that I have subsequently worked with. My two-volume "Unity of Forces in the Universe" was produced in 1981, literally a month or so after I submitted the manuscript. According to Amazon.com, it came out on January 1, 1982, less than a year and a month after the company was registered. I was told that Unity was one of the first bestsellers WSP published in its early years, among the top five in sales.

The enormous success of WSP over the decades is due to many factors, most notably the forceful personality and vision of K.K. Phua, but one of which is certainly speedy publication on timely topics of urgent interest.

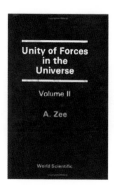

Unity of Forces in the Universe (2 Volume Set)
by Anthony Zee | Jan 1, 1982

Hardcover

$225^{00}

FREE Delivery for Prime members
Temporarily out of stock.

More Buying Choices
$169.02 (3 used & new offers)

Fig. 1.   Amazon now wants $225 for the hard cover.

That *Unity* contained reprinted papers partly accounted for the phenomenal speed of publication; nevertheless, I took the task of introducing each section to the reader quite seriously. For years, people even urged me to collect my various introductions, such as the one to grand unified theory, to cosmology, and to group theory, and publish them separately as a book. While the younger generations of physicists in the era of Google Scholar and the physics archive and laser printers are justifiably puzzled by these reprint volumes, there was a time when they served an essential purpose in physics. I learned about various research areas, such as SU (3) and current algebra, by studying reprint volumes. By the late 1970s, the particle theory community clearly needed a reprint volume on grand unified theory.

Unity, at a 1000+ pages, has sold close to 2000 copies. In addition, WSP gave away an untold number of complimentary copies. For decades, I would see — and I still see — my two black volumes on the shelves of many particle theorists. A few years ago, I was touched when a leading particle phenomenologist wrote to me to say that, while he, like many others, wanted to thank me for my trilogy of textbooks, the book he really wanted to thank me for was "Unity of Forces in the Universe". Having all those classic papers at his fingertips was invaluable during his student days and his subsequent research career.

K.K. Phua has played several important roles in physics over the last four decades. First of all, by establishing World Scientific he gave voice to authors from various Asian countries, and later, from the Soviet Union and

Eastern European countries, who would otherwise have difficulty getting their books published in the western world. As WSP grew and became better known, KK was able to persuade a number of highly distinguished Nobel laureates and near Nobel-level physicists, notably Phil Anderson among many others, to publish with WSP. There is no question that World Scientific books now play a prominent role in physics libraries.

Secondly, by organizing an enormous number of conferences held in Singapore, he has brought an awareness of the modern society that is Singapore to the physics community. Furthermore, many physicists who might not otherwise have crossed paths regularly were brought together periodically in Singapore. Here, for example, is a photo taken at Murray Gell-Mann 80th birthday.

Fig. 2.   At Murray Gell-Mann 80th birthday: K.K. Phua and I stood next to each other.

Thirdly, the proceedings of these conferences published by WSP serve to record various periods in physics. I have many such volumes in my office and at home.

Fourthly, at many of these conferences, students from developing countries and from Singapore have the opportunities to meet some of the leading physicists that they might not otherwise have met. I for one have enjoyed talking with some of these students.

It is truly remarkable how much WSP has grown from its beginning in late 1980. I am told that it now publishes almost 600 books a year and has 150 journals covering physics, mathematics, computer science, engineering, life sciences, medical sciences, social sciences, economics and finance. All of this testify to K.K. Phua's clout and influence.

Acknowledgment

I would like thank Lakshmi Narayanan for providing me with various facts about WSP and about my book.

Short Tributes

生
日
快
乐

Dear Dr. Phua,

Happy 80th! In the Indian tradition, it is an especially auspicious occasion to look back and assess. In your case, there are so many things to celebrate. You were a true pioneer to burst into the scientific world with very original ideas on scientific publishing. They have made a huge impact across the world, and across generations of physicists and mathematicians.

It must be a very satisfying feeling!

I recall our very first encounter when you approached me to publish the monograph "New Perspectives on Canonical Gravity" some 30 years ago. I was moved by your motivation to help Prof. Fang Lizhi who had just left China and was in somewhat of a rough patch in his life. Then, over 10 years ago you generously contributed to the International Society on General Relativity and Gravitation to award S. Chandrasekhar prizes for best postdoc presentations in their triennial conferences. World Scientific is recognized with much appreciation as the benefactor of this award. Perhaps more importantly, several of the postdocs who won that prize have become influential researchers and are emerging as future leaders. On a more personal level, I would like to express my heartfelt appreciation for several World Scientific books you sent me over the years, on your own. The one by Freeman Dyson and the one documenting Chandrasekhar's early correspondence, in particular, have become prominent features of my library — I found them tremendously inspiring and I often recommend them to bright young researchers.

I look forward to years of happy collaboration with you on various World Scientific projects, such as the 100 Years of General Relativity Series.

Best wishes,

Abhay Ashtekar
Founding Director,
Institute for Gravitation & the Cosmos
Penn State, University Park, PA 1680, USA

Dear Kok Khoo,

I am greatly impressed by your achievements in organizing WSPC and series of conferences. At our first meeting at ISMD-1978 in Volendam, we first discussed your two-component model and then you started to tell me about your "secret private plans" to found and operate the new Publishing House. How young were we! In 1990, this project was successfully evolving as you told me and my wife, inviting us to a cup of coffee at the English Club in Singapore when I participated in Rochester-type conference organized by you. Afterwards, I actively contributed to WSPC journals *IJMPA* and *MPLA* both as an editor and as an author. My latest article is being published just now. Also I have edited the English translation of E. L. Feinberg's book of reminiscenses on prominent physicists. Your skill in operating WSPC is clearly seen in the effective and accurate work of the whole staff and the editorial boards. My congratulations on your anniversary and many wishes for good health, happiness and further successes!

Igor Dremin
Lebedev Physical Institute, Moscow, Russia

Dear Professor KK Phua,

As all members of our scientific community, I would like to praise your remarkable and pioneering action in founding and running World Scientific. This is a very impressive achievement that has and will continue having a long-lasting impact on our scientific work.

I wish you a Happy Birthday and many more years of health and happiness.

Marc Henneaux
Director
International Solvay Institutes for Physics and Chemistry, Belgium

Dear Professor KK Phua,

It has always been a great pleasure and honor to collaborate with you on various projects, from your excellent journals to my books published by World Scientific to Proceedings of conferences which I was organizing. I admire your kind and attentive approach to authors and editors, your willingness to do things. It was great to see you in person in Singapore back in 1990 at the XXV International Conference on High Energy Physics, an outstanding event organized by you and your team. I wish you strong health, happiness and success for many years ahead.

Valery Rubakov
Russian Academy of Sciences

Dear Professor KK Phua,

I have greatly admired your foresight and skill in founding and operating World Scientific. I have also enjoyed meeting you in person during my visits to Singapore. I wish you many more years of health and happiness.

John Schwarz
Harold Brown Professor of Theoretical Physics, Emeritus
Caltech

Over the years World Scientific Publishing Co. organized by Dr. Phua produced a significant impact on the research community and for many became the Publisher of choice. The basic ideas and principles laid in its foundation — in particular, accessibility — proved to be highly successful. I wish Dr. Phua to continue to shape the global market of scientific publications.

Mikhail Shifman
University of Michigan, USA

Part II

Research Papers

Do We Know How to Define Energy in General Relativity?

Sinya Aoki

Center for Gravitational Physics and Quantum Information,
Yukawa Institute for Theoretical Physics, Kyoto University, Kitashirakawa Oiwakecho,
Sakyo-ku, Kyoto 606-8502, Japan
saoki@yukawa.kyoto-u.ac.jp

This essay is dedicated to Prof. K.K. Phua on the occasion of his 80th birthday. While the contents of this essay are based on our recent papers [1,2] published in the *International Journal of Modern Physics A*, I have added many personal opinions, so that I am solely responsible for all statements in this essay.

1. Introduction

You may think that an answer to the question in the title of this essay must be "Yes", since a concept of energy is essentially important in physics so that it can be well understood in general relativity, which has now matured more than 100 years after Einstein's first proposal in 1915 [3]. As I will show in this essay, however, the correct answer is probably "No" or more precisely "A concept of energy is not well understood in general relativity.".

Let me start with my discussion on Einstein equation in general relativity, which reads

$$R^{\mu\nu} - \frac{1}{2}g^{\mu\nu}R = 8\pi G T^{\mu\nu} \tag{1}$$

where $T^{\mu\nu}$ on the right-hand side is an energy momentum tensor (EMT) of matter, while the Ricci tensor $R^{\mu\nu}$, the Ricci scalar R and the metric $g^{\mu\nu}$ on the left-hand side describe how a spacetime is curved, and G is the Newton constant, which controls a coupling strength of matter with spacetime. The

Newton constant is very small as $G = 6.7 \times 10^{-39}$ (GeV)$^{-2}$ in Natural unit ($c = \hbar = 1$). The Einstein equation tells us how existing matter determines the structure of curved spacetime.

Einstein derived this equation from two assumptions — the invariance under the general coordinate transformation (general covariance) and the equivalence principle. The former means that the law of gravity in physics does not depend on a choice of a coordinate system that describes spacetime, while the latter says that a gravitational force can be locally removed away by the coordinate transformation.

2. Dilemma: Conserved Energy or Coordinate Independence?

A covariant derivative applied to the left-hand side of the Einstein equation vanishes identically. This is a consequence of the *Bianchi identity*, which implies the covariant conservation of the EMT as

$$\nabla_\mu T^{\mu\nu} := \partial_\mu T^{\mu\nu} + \Gamma^\mu_{\mu\alpha} T^{\alpha\nu} + \Gamma^\nu_{\mu\alpha} T^{\mu\alpha} = 0, \tag{2}$$

as a generalization of that in flat spacetime, $\partial_\mu T^{\mu\nu} = 0$, where $\Gamma^\nu_{\mu\alpha}$ is called a connection or a Christoffel symbol. Generally, Eq. (2) does not identically hold but is satisfied only after equations of motion for matter are used. While a conservation of energy in flat spacetime is derived from $\partial_\mu T^{\mu\nu} = 0$, Eq. (2) does not provide a conserved energy in general, due to extra second and third contributions in the covariant derivative. Then, what is conserved energy in general relativity?

One of the textbook answers was proposed by Einstein himself. He modified the EMT as $\tilde{T}^{\mu\nu} := T^{\mu\nu} + t^{\mu\nu}$, which satisfies

$$\partial_\mu(\sqrt{-g}\tilde{T}^{\mu\nu}) = 0, \tag{3}$$

where g is a determinant of $g_{\mu\nu}$ and $\sqrt{-g}$ is a spacetime volume density. Thus a conserved energy may be defined as

$$E = \int d^3x \sqrt{-g}\tilde{T}_{00} \quad \text{or} \quad E = -\int d^3x \sqrt{-g}\tilde{T}^0{}_0, \tag{4}$$

where an integral measure $\sqrt{-g}d^3x$ is covariant under general coordinate transformations. Unfortunately there exists no theoretical preference between $\widetilde{T}_{00}$ and $\widetilde{T}^0{}_0$ in this approach. A problem of Einsteins's such approach is that $t^{\mu\nu}$ is not a tensor under general coordinate transformations, as evident from that Eq. (3) is not a covariant equation. Thus $\widetilde{T}^{\mu\nu}$ does not satisfy one of the two fundamental assumptions in general relativity, the general covariance, and $\widetilde{T}^{\mu\nu}$ is called Einstein's pseudo-tensor due to this violation.

There is a dilemma about the choice between non-conservation of energy or the violation of general covariance. In my opinion, Einstein made the worse choice, but after some debates on its limitations and problems, the pseudo-tensor approach has been gradually accepted for the definition of energy in general relativity, partly based on arguments that $t^{\mu\nu}$ represents the energy of gravitational fields, which hence must be non-covariant and coordinate dependent since the equivalence principle always leads to gravitational fields that vanish by a coordinate transformation. I feel that this view, found frequently in some textbooks, is not scientifically convincing to guarantee the correctness of the pseudo-tensor, since we may invert the logic that we disfavor the pseudo-tensor since we can always make $t_{\mu\nu}$ zero locally.

Later, there appeared conserved and covariant definitions, called quasi-local energies, which gives a total energy only as surface integrals without knowing a local energy density, and include the well-known ADM energy [4]. Quasi-local energies, however, cannot solve our dilemma, since their conservation is merely an identity as a consequence of general coordinate transformations, known as the Noether's 2nd theorem [5], and this flaw is also true for the pseudo-tensor [6]. Under such unsatisfactory circumstances for a definition of energy in general relativity, we have recently proposed an alternative, which will be explained in this essay.

3. Alternative Definition of Energy in General Relativity

In Ref. [1], we consider a case that a metric allows a time-like or stationary Killing vector satisfying $\nabla_\mu \xi_\nu + \nabla_\nu \xi_\mu = 0$. For example, if a metric is independent of a time coordinate x^0, the stationary Killing vector is given by $\xi^\mu = \xi^\mu_0 := -\delta^\mu_0$, where a minus sign is chosen for energy, given later, positive. In this case, we have proposed an alternative definition of energy in general relativity as

$$E_{ours} = \int_{\Sigma} [d^3x]_{\mu} \sqrt{-g} T^{\mu}{}_{\nu} \xi^{\nu}, \tag{5}$$

that is conserved since $\partial_{\mu}(\sqrt{-g} T^{\mu}{}_{\nu} \xi^{\nu}) = \sqrt{-g} \nabla_{\mu}(T^{\mu}{}_{\nu} \xi^{\nu}) = \sqrt{-g} T^{\mu\nu}(\nabla_{\mu}\xi_{\nu} + \nabla_{\nu}\xi_{\mu})/2 = 0$, where Σ is a space-like entire hyper-surface with a volume factor $[d^3x]_{\mu}$. Since Eq. (5) is manifestly invariant under the general coordinate transformation, we confidently say "Yes" to the question in the title of this essay in special cases while avoiding the dilemma. Unlike the pseudo-tensor and quasi-local energies, Eq. (5) passes a minimum test for energy in general relativity to satisfy that it leads to the standard energy in flat spacetime in the Newton constant $G \to 0$ limit as

$$\lim_{G \to 0} E_{ours} = -\int_{x^0=fix} d^3x \, T^0{}_0 = \int_{x^0=fix} d^3x \, T_{00}. \tag{6}$$

While Eq. (5) or similar ones can be found in some textbooks, there were no applications of the definition before we wrote a paper [1], where several examples were considered. As a simple example, an energy of the Schwarzschild blackhole becomes

$$E_{ours}^{BH} = \int d^3x \, \sqrt{-g} T^0{}_{\nu} \xi_0^{\nu} = M, \tag{7}$$

where the EMT of the blackhole is given as

$$T^0{}_0 = -\frac{1}{4\pi} \frac{\partial_r(M\theta(r))}{r^2} = -\frac{1}{4\pi} \frac{M\delta(r)}{r^2}, \tag{8}$$

which shows that the energy is concentrated at the origin producing singularity. Thus, unlike a folklore that the Schwarzschild blackhole is a vacuum solution to the Einstein equation (1), it has the non-zero EMT at the origin so that the Schwarzschild blackhole is indeed NOT a vacuum solution. Since only a local property near $r = 0$ is relevant for this derivation, a mechanism for non-zero EMT at $r = 0$ is identical to that which produces a δ function singularity for the point charge in the Maxwell equation. As I have been rather surprised, however, many people including those who are regarded to have the highest intelligence in the world claim or believe that the Schwarzschild blackhole is a vacuum solution to the Einstein equation, even though physically a gravitational collapse of matter leads to a black holes as a final state.[a] I suspect that this is the reason why the formula (5) has not been used so far, though the definition has been known for a long time: If the Schwarzschild blackhole were a vacuum solution, the definition (5) would give zero, an unacceptable result for blackhole energy.

4. Energy of a Neutron Star

In the case of the Schwarzschild blackhole, there appears no difference between the energy given in (5) and the pseudo-tensor/quasi-local energy [1,6]. We here consider energy of a compact star such as a neutron star, which reveals a difference between the two [1].

Energy defined by (5) for a static and spherically symmetric compact star becomes

$$E_{our} := -\int d^3x \sqrt{-g}T^0{}_0 = 4\pi \int_0^R dr \sqrt{-g_{00}(r)g_{rr}(r)}r^2\rho(r), \qquad (9)$$

where R is a radius of the compact star, $g_{00}(r)$ and $g_{rr}(r)$ are time and radial components of a diagonal metric, respectively, and $\rho(r)$ is a matter density related to the EMT as $T^0{}_0 = -\rho(r)$. On the other hand, the ADM energy, called Misner–Sharp mass, is given as

$$E_{ADM} = 4\pi \int_0^R dr\, r^2\rho(r), \qquad (10)$$

which does not have the $\sqrt{-g}$ factor, and thus is not a covariant. Note that, since $g_{00}(r)g_{rr}(r) = -1$ for the Schwarzschild blackhole, both definitions agree.

Let us compare these two energies in the Newtonian limit that G is small and $P(r) \ll \rho(r)$ where $P(r)$ is a pressure of the matter. In this limit we

---

[a]It is often claimed that singularities of black holes in general relativity cannot be described by distributions such as δ function due to nonlinearities of the theory. As shown in (8), however, the singularity of the Schwarzschild blackhole indeed leads to δ function singularity in the EMT without products of distributions. Mathematically this means that the Einstein equation (1) should be understood in a distributional sense [7]. A non-vanishing EMT has been calculated also for the Kerr blackhole [8].

have

$$E_{\text{our}} - E_{\text{ADM}} = \frac{G}{2} \int d^3x \rho_0(|\vec{x}|)\phi(\vec{x}) + O(G^2, G\omega), \quad \omega := \max_r \frac{P(r)}{\rho(r)}, \quad (11)$$

where $\rho_0(|\vec{x}|)$ is the matter density at $\vec{x}$ in this limit and $\phi(\vec{x})$ is the gravitational potential at the point. On the other hand, E_{ADM} in the Newtonian limit becomes

$$E_{\text{ADM}} = \int d^3x \, \rho_0(|\vec{x}|) + \frac{G}{2} \int d^3x \rho_0(|\vec{x}|)\phi(\vec{x}) + O(G^2, G\omega). \quad (12)$$

Thus E_{our} represents the energy of matter in the presence of the background gravitational potential $\phi(\vec{x})$, while E_{ADM} is the energy of the self-interacting matter,[b] where

$$\phi(\vec{x}) = -\int d^3y \frac{\rho_0(|\vec{y}|)}{|\vec{x} - \vec{y}|}. \quad (13)$$

Therefore both seem reasonable for different situations.[c] It will be interesting to understand the meaning of the difference.

5. Generalization

The energy given in (5) is not the final answer, since a metric usually does not have a Killing vector ξ_0^ν in general. Then, how can we define energy for a general metric in the absence of the Killing vector?

While ξ_0^μ is no more a Killing vector for a general metric, the energy $E[\xi_0]$ in (5) is still conserved if the metric satisfies [2],

$$T^\mu{}_\nu \nabla_\mu \xi_0^\nu = -T^\mu{}_\nu \Gamma^\nu{}_{\mu 0} = 0, \quad (14)$$

where summations over μ, ν are implicitly assumed. The conservation of $E[\xi_0]$ is easily confirmed, since

$$\partial_\mu \left(\sqrt{-g} T^\mu{}_\nu \xi_0^\nu \right) = \sqrt{-g} \nabla_\mu (T^\mu{}_\nu \xi_0^\nu) = \sqrt{-g} T^\mu{}_\nu \nabla_\mu \xi_0^\nu = 0, \quad (15)$$

---

[b]I would like to thank Prof. Masaru Shibata for pointing out this difference to me.
[c]Note however that this conclusion is different from the one in Ref. [1]. Thus I am solely responsible for the content in this section.

where we use $\nabla_\mu T^\mu{}_\nu = 0$ and a property of the covariant derivative $\sqrt{-g}\nabla_\mu J^\mu = \partial_\mu(\sqrt{-g}J^\mu)$ for an arbitrary vector J^μ. For example, the energy is indeed conserved during some types of gravitational collapses [2].

For more general cases, however, $E[\xi_0]$ is no more conserved. A simple example is a model of a homogeneous and isotropic expanding universe with the Friedmann–Lemaître–Robertson–Walker (FLRW) metric, given by

$$ds^2 = -(dx^0)^2 + a^2(x^0)d\vec{x}^2, \tag{16}$$

where $a(x^0)$ is a scale factor. The EMT is given by the perfect fluid as

$$T^0{}_0 = -\rho(x^0), \quad T^i{}_j(x^0) = P(x^0)\delta^i_j, \quad T^0{}_j = T^i{}_0 = 0, \tag{17}$$

whose conservation $\nabla_\mu T^\mu{}_\nu = 0$ implies

$$\dot{\rho} + 3(\rho + P)\frac{\dot{a}}{a} = 0, \tag{18}$$

where $\dot{f}$ means a derivative of f with respect to a time coordinate x^0.

The energy (5) is easily calculated as

$$E := E[\xi_0] = -\int d^3x \sqrt{-g}T^0{}_0 = V_0 a^3 \rho, \tag{19}$$

where $V_0 := \int d^3x$ is a constant space volume without the scale factor $a(x^0)$. This energy is indeed not conserved as

$$\dot{E} = V_0(\dot{\rho}a^3 + 3\rho\dot{a}a^2) = -3\frac{\dot{a}}{a}\frac{P}{\rho}E \neq 0, \tag{20}$$

as long as $P \neq 0$ and $\dot{a} \neq 0$.

In Ref. [2], we have proposed a method to define a conserved quantity as a generalization of the energy, which is given by

$$S := E[\beta\xi_0] = -\int d^3x \sqrt{-g}T^0{}_0\beta = \beta E, \tag{21}$$

where $\zeta^\mu(x^0) := \beta(x^0)\xi^\mu_0$ should satisfy

$$T^\mu{}_\nu \nabla_\mu \zeta^\nu \left(= \rho\dot{\beta} - 3P\frac{\dot{a}}{a}\beta \right) = 0. \tag{22}$$

Then it is easy to see that S is conserved as

$$\dot{S} = \beta \dot{E} + \dot{\beta} E = -3\frac{\dot{a}}{a}\frac{P}{\rho}\beta E + 3\frac{\dot{a}}{a}\frac{P}{\rho}\beta E = 0. \tag{23}$$

What is the physical meaning of S? Using (22), we rewrite the above conservation equation as

$$\frac{dS}{dx^0} = \frac{dE}{dx^0}\beta + E\frac{d\beta}{dx^0} = \left(\frac{dE}{dx^0} + P\frac{dV}{dx^0}\right)\beta, \tag{24}$$

where $V(x^0) := V_0 a^3(x^0)$ is a time-dependent space volume during the expansion. Since (24) is similar to the first law of thermodynamics,

$$T dS = dE + P dV, \tag{25}$$

we interpret S as the total *entropy* of the universe and $\beta = \dfrac{1}{T}$ as an inverse temperature. Thus, although the total energy is not conserved, the total entropy is conserved during the expansion. In addition, $\beta(x^0)$ gives the time-dependent inverse temperature of the universe, which increases (equivalently the temperature decreases) as the universe expands, since

$$\frac{\dot{\beta}}{\beta} = 3\frac{P}{\rho}\frac{\dot{a}}{a} > 0. \tag{26}$$

Even for more general cases than the FLRW universe, using a vector ζ^μ satisfying (22), we can always construct a conserved charge as $S := E[\zeta]$, which can be interpreted as an entropy in some cases [2], but whose physical interpretation for a general spacetime is still missing. It will be interesting and important not only to find a more general physical meaning of S than the entropy but also to understand a mechanism working behind this conservation law in general relativity.

6. Do Gravitational Fields Carry Energies?

As a concluding remark, I consider the above question. Since our definition of the energy or its generalization requires a non-zero EMT, gravitational fields including gravitational waves cannot have such a conserved quantity.

This seems to contradict the majority's common sense that gravitational waves carry energies. After careful consideration, however, we realize that there exists no theoretical as well as experimental evidence which unambiguously prove that gravitational waves carry energies,[d] as I had already discussed that a concept of energy for gravitational fields is not easily justified in general relativity.

Let me be more conservative. Logically it is certainly possible to define another quantity associated with gravitational fields, which may be regarded as gravitational energy. Therefore, somebody might find such a new definition of gravitational energy in the future, which at the same time, is exchangeable between matter and gravitational fields, so that only the sum of matter energy in (5) and the gravitational energy is conserved for general spacetime. Even though such a possibility cannot be ruled out so far, we have established in our paper [2] that there *always* exists a conserved quantity defined by $E[\zeta]$ with ζ satisfying (22), as carried only by matter.

Let me go back to the start of this essay. The Einstein equation (1) clearly tells us a matter $T^{\mu\nu}$ generates a gravitational field. On the other hand, the gravitational field itself generates no additional gravitational field. This structure of the Einstein equation seems naturally to explain why gravitational fields carry no energy.

Finally, it is my belief that the definition of energy in general relativity is not completely understood yet, as is evidenced by the existence of various proposals for energy in literature. It is hoped that the present essay will trigger further analyses of this interesting issue.

Acknowledgments

I have enjoyed fruitful collaborations with Drs. Tetsuya Onogi and Shuichi Yokoyama. Without them, I could never have written this essay.

---

[d]It is often said that an orbital decay for a binary star system is explained by the loss of their energy due to gravitational waves. More precisely, however, it is confirmed only that a rate of the orbital decay is consistent with the Einstein equation in some approximation, where a description of the gravitational wave energy is valid within the approximation.

References

[1] S. Aoki, T. Onogi and S. Yokoyama, *Int. J. Mod. Phys. A* **36** (2021) 2150098 doi:10.1142/S0217751X21500986 [arXiv:2005.13233 [gr-qc]].

[2] S. Aoki, T. Onogi and S. Yokoyama, *Int. J. Mod. Phys. A* **36** (2021) 2150201 doi:10.1142/S0217751X21502018 [arXiv:2010.07660 [gr-qc]].

[3] A. Einstein, *Ann. der. Phys. Ser.* 4, **49** (1916), pp. 769–822.

[4] R. L. Arnowitt, S. Deser and C. W. Misner, in *Gravitation: An Introduction to Current Research*, L. Witten, ed. (Wiley, New York, 1962). See also *Gen. Rel. Grav.* **40** (2008) 1997–2027 doi:10.1007/s10714-008-0661-1.

[5] E. Noether, *Gott. Nachr.* **1918** (1918) 235–257 doi:10.1080/00411457108231446 [arXiv:physics/0503066 [physics]].

[6] S. Aoki and T. Onogi, Conserved non-Noether charge in general relativity: Physical definition vs. Noether's 2nd theorem [arXiv:2201.09557 [hep-th]].

[7] H. Balasin and H. Nachbagauer, *Class. Quant. Grav.* **10** (1993) 2271 doi: 10.1088/0264-9381/10/11/010 [arXiv:gr-qc/9305009 [gr-qc]].

[8] H. Balasin and H. Nachbagauer, *Class. Quant. Grav.* **11** (1994) 1453–1462 doi: 10.1088/0264-9381/11/6/010 [arXiv:gr-qc/9312028 [gr-qc]].

Quantum Computations and Option Pricing

生
日
快
乐

Belal Ehsan Baaquie

Helixtap Technologies,
75 Ayer Rajah Crescent, #01-02, Singapore 139953
belal@helixtap.com

Option pricing is briefly reviewed. A quantum algorithm that is equivalent to classical Monte Carlo algorithms is discussed. The algorithm for a quadratic improvement compared to the Monte Carlo error is then analyzed in some detail. The derivation is seen to use quantum Fourier transforms as well as the technique of amplitude amplification that is similar to the Grover search algorithm.

1. Dr. K.K. Phua

My first meeting with Dr. K.K. Phua was in 1982 at ICTP, Trieste. We were introduced by the late Professor Abdus Salam, who was my mentor and a close friend and colleague of Dr. Phua. In 1983, Dr. Phua kindly invited me to Singapore to attend the First Asia-Pacific Conference in Physics and it was also my first visit to the Garden City. I joined the Department of Physics, National University of Singapore in 1984, where Dr. Phua was also a faculty member and I was warmly welcomed by him to the department, and to Singapore.

For the next 30 years, I worked with Dr. Phua on many projects of common interest including organizing numerous conferences and meetings. I was one of the Editors of the first journal launched by World Scientific in 1985 and continued my fruitful collaboration with Dr. Phua for launching other journals as well. I have published many papers in the journals of World

Scientific as well as a few books, and their team has always done an excellent job.

It has been a pleasure and privilege to have worked closely with Dr. Phua and to have become one of his friends. I have always been impressed by Dr. Phua's visionary outlook, his drive towards excellence as well as his positive and optimistic approach to all the challenges that he has faced. Dr. Phua, along with his wife Ms. Doreen Liu, are the driving force that has made World Scientific grow, from being a start-up in 1981, into a world class publisher — with hundreds of books and journals being published every year. The growth of World Scientific is a tribute to Dr. Phua's vision of creating a world class publishing house that is firmly rooted in Asia, and serving the needs of the international scientific community as well as that of the developing countries.

Dr. Phua has always been a source of great support to all his friends, colleagues and co-workers, including myself, and his generous and kind personality has won him many friends the world over. On this auspicious occasion of his 80th birthday, I wish Dr. Phua a long, happy and healthy life, full of even greater achievements, and look forward to continuing our friendship and collaboration.

2. Quantum Algorithms and Finance

Quantum theory introduces a fundamentally new framework for thinking about Nature and entails a radical break from the paradigm of classical physics. In spite of the fact that the *shift of paradigm* from classical to quantum mechanics has been ongoing for more than a century, a complete conceptual grasp of quantum mechanics has till today proved elusive. According to leading quantum theorist Richard Feynman, *"It is safe to say that no one understands quantum mechanics"*.[1]

The idea of quantum computers was pioneered by Yuri Manin (1980) and Richard Feynman (1981), when they discussed simulations that could not be carried out by a classical computer — but only by a quantum computer. Feynman discussed the ideas of quantum gates and quantum computers.[2] In 1983, David Deutsch made a crucial advance by proposing that the quantum qubit is the natural generalization of the bit of classical computers.[3] The field over the last two decades has seen phenomenal growth and the development of a practical quantum computer is poised to be game-changing technology of the 21st century. A survey of quantum computers has been discussed by Baaquie[1] and this article is largely based on an introductory text of this field by Baaquie and Kwek.[4]

Classical computers refer to computers that are constructed with an electronics circuit based on the existing paradigm of classical computer science, and the design of hardware is based on classical physics. [The components of a classical computer may be based on quantum mechanics, such as semiconductors, but these components function as classical devices in a classical computer.] Quantum computers, in contrast, are devices based on the principles of quantum mechanics. The discipline of quantum computers and quantum computing is an emerging and rapidly developing subject.

Quantum algorithms are directly based on the principles of quantum mechanics and run on quantum computers; these are codes that can vastly speed-up information processing, far outpacing currently used classical computers. Quantum computers have potential applications in a vast number of fields, such as cryptography, manufacturing, information and communications technology, finance, autonomous vehicles, pharmaceuticals, energy, logistics, artificial intelligence and so on. In recent years, there is hardly a day that goes by without some articles touching on the potential importance of quantum computers, as well as on the intense competition among various companies and countries that are trying to take the lead in developing a practical quantum computing device.

It is the view of many experts that the breakthrough for quantum algorithms is going to take place in finance. Unlike in devices such as solar panels where the algorithmic content is negligible compared to the hardware component, in finance, efficient algorithms can have a far greater impact. The reason is that even the slightest improvement in the performance of a quantum algorithm over the classical algorithm can lead to substantial gains in finance, given the massive volume of debt in capital markets, with global

---

[1]https://futureoffinance.biz/2021/07/13/where-quantum-computers-are-now-and-where-they-are-going-next/

debt reaching \$226 trillion by the end of 2020[2]; the options market has a notional value of over US\$650 trillion (with actual value of US\$12 trillion).

To develop efficient financial models for this trillion dollar market — that in particular could provide solutions in real-time — needs enormous computing power that would greatly benefit from a practical quantum computer. Furthermore, due to the financial markets constantly providing up-to-date data, quantum algorithms can be calibrated and tested far more efficiently than for hardware-oriented devices that are based on quantum computers.

For reasons stated above, the application of quantum algorithms to pricing options is briefly discussed. The approach of this paper is the one adopted by Baaquie:[5–7] the principles of finance are taken from the existing theories of finance. Financial instruments are not assumed to have any quantum indeterminacy — but rather, the uncertainty in the future values of financial instruments are determined by classical stochastic calculus. The quantum algorithm discussed provides an efficient procedure for evaluating the option price of a European call or a put option. Other more complex financial instruments, especially debt instruments, are left for further reading.

3. Review of Option Pricing

The price of all options is the discounted value of the expectation value of a payoff function in the future, with the future uncertain price of financial instruments like stocks or bonds being represented by classical random variables. Most numerical studies of option pricing are based either on solving the relevant partial differential equations or on Monte Carlo simulations of the stochastic equations.[8]

Consider the price of a path independent option $C(x,t,T)$ at present time t, and which matures at future time T. The payoff function of the option $v(x)$ is defined to be the price of the option when it matures at future time T and is given by

$$C(x,T,T) = e^{-r\tau}v(x)$$

---

[2]https://www.visualcapitalist.com/global-debt-to-gdp-ratio/

The payoff function of a call option is given as

$$v(x) = [e^x - K]_+$$

Let the initial stock price be $S = e^{x_0}$, with spot interest rate given as r and the remaining time for maturity of option given by $\tau = T - t$ and let the price of the option be $C(x, \tau)$. Let $P(x_0, x; \tau)$ be the conditional probability for which the stock price is $S = e^{x_0}$ today, it has a value of e^x after time τ; the discounted price of the payoff yields the option price, and which is given by

$$C(x_0, \tau) = e^{-r\tau} \int dx P(x_0, x; \tau)v(x) = e^{-r\tau} \int dx p(x)v(x) = e^{-r\tau} E[v] \quad (1)$$

where

$$p(x) = P(x_0, x; \tau) \quad (2)$$

Ignoring for the rest of the discussion the discounting by the spot interest rate, the evaluation of the call option is given by evaluating the expectation value $E[v]$. Equation (1) yields the following expectation value

$$E[v] = \int dx p(x)v(x) \quad (3)$$

Note that a European and American call option's price is always less than the security $S = e^{x_0}$;[8] hence by dividing out by $S = e^{x_0}$, one can always redefine the payoff function so that

$$\mu = E[e^{-x_0}v] < 1 \quad (4)$$

For ease of notation, we use v and $e^{-x_0}v$ interchangeably, and the difference will be known from the context of the equation.

If one does a Monte Carlo simulation for the evolution of the stock price using its stochastic differential equation, then after N trials of evolving the initial value of the log of stock x_0 to its final values $x^{(i)}$, the Monte Carlo estimate of the expectation value, from Eq. (3), is given as

$$E_{MC}[v] = \frac{1}{N} \sum_{i=1}^{N} x^{(i)} \pm \frac{\sigma_v}{\sqrt{N}} \quad \text{with 66\% likelihood}$$

The quantum algorithm, after amplitude amplification, gives a *quadratic improvement* and yields

$$E[v] = \sum_{i=0}^{N-1} p(x^{(i)})v(x^{(i)}) \pm \frac{\sigma_v}{N} \text{ with 66\% likelihood}$$

In simplified notation, the summation is rewritten as

$$E[v] = \sum_{x=0}^{N-1} p(x)v(x) \pm \frac{\sigma_v}{N} \text{ with 66\% likelihood} \tag{5}$$

This improvement by a factor of a square root is similar to the improvement of the Grover algorithm and the quantum algorithm discussed is closely related to the Grover algorithm.[4]

4. Quantum Algorithm

The basic strategy for applying quantum algorithms to option pricing is to map the computation required for evaluating the option's price to an equivalent quantum algorithm. One is then free to use the resources of both quantum entanglement and superposition in improving the efficiency of the quantum algorithm, which for option pricing achieves a quadratic improvement when compared to the classical Monte Carlo simulation. It is worth noting that one is evaluating a classical quantity using the computational tools of a quantum algorithm.

A quantum algorithm for pricing options has the following distinct steps.[4]

(1) The classical random variable x with the probability distribution function $p(x)$ given by Eq. (2) is discretized and replaced by n binary quantum degrees of freedom which yields the basis states $|x\rangle$ with $x = 0, 1, \cdots, 2^n - 1$, with $N = 2^n$.

(2) The basis states are combined into a superposed state to encode the option's data.

(3) The probability distribution function $p(x)$ defines the function $a(x) = \sqrt{p(x)}$ and the initial data of the option is encoded into the state vector $|\psi\rangle$.

(4) The superposed basis states $|x\rangle$ yield the n qubits state vector

$$|\psi\rangle = \sum_{x=0}^{2^n-1} a(x)|x\rangle \; ; \; \sum_{x=0}^{2^n-1} |a(x)|^2 = 1$$

(5) An entangled state that entangles the state vector $|\psi\rangle$ with the payoff function v is created using an auxiliary qubit and is given by

$$|\psi\rangle \to |\eta\rangle = \sum_{x=0}^{2^n-1} a(x)|x\rangle\left(\sqrt{1-v(x)}|0\rangle + \sqrt{v(x)}|1\rangle\right)$$

(6) In Sec. 6 onwards, amplitude amplification is applied to the quantum state $|\eta\rangle$. A quantum gate (oracle) is applied simultaneously to all the basis states and the auxiliary qubit. At the end of the application of the gate, one of the states that holds the correct answer is marked, as in the case for Grover's algorithm.

(7) The probability of measuring the marked state is amplified using quantum gates that achieve the amplification using quantum interference.

(8) A generalized Born measurement is carried out on the final qubits.

5. Quantum Algorithm for Expectation Value

We discuss a quantum algorithm that is valid for any expectation value. The algorithm for amplitude amplification is built on this algorithm.

The required quantum algorithm yields the following expectation value

$$E[v] = \sum_{x=0}^{2^n-1} |a(x)|^2 v(x)$$

The algorithm is defined by the following.[9] Consider an ancilla qubit state $|0\rangle$ and a rotation R such that

$$R\left(|x\rangle|0\rangle\right) = |x\rangle\left(\sqrt{1-v(x)}|0\rangle + \sqrt{v(x)}|1\rangle\right)$$

The algorithm $\mathcal{A}$ acts on the input qubits state and yields

$$\mathcal{A}|0\rangle^{\otimes n} = \sum_{x=0}^{2^n-1} a(x)|x\rangle \tag{6}$$

Hence

$$|\eta\rangle = R\left(\sum_{x=0}^{2^n-1} a(x)|x\rangle|0\rangle\right) = \sum_{x=0}^{2^n-1} a(x)|x\rangle\left(\sqrt{1-v(x)}|0\rangle + \sqrt{v(x)}|1\rangle\right)$$

$$\Rightarrow |\eta\rangle = \sum_{x=0}^{2^n-1} a(x)|x\rangle\left(\sqrt{1-v(x)}|0\rangle + \sqrt{v(x)}|1\rangle\right) \quad : \quad \langle\eta|\eta\rangle = 1 \qquad (7)$$

The state vector $|\eta\rangle$ is a rotation of state vector $|\psi\rangle$ through an angle γ as shown in Fig. 1. The unitary transformation $\mathcal{F}$ is given by

$$\mathcal{F} = R(\mathcal{A} \otimes \mathbb{I}_2) \quad : \quad \mathcal{F}\left(|0\rangle^{\otimes n}|0\rangle\right) = \left(R(\mathcal{A} \otimes \mathbb{I}_2)\right)\left(|0\rangle^{\otimes n}|0\rangle\right) = |\eta\rangle \qquad (8)$$

Note no amplification of any of the states was carried out. In preparation for a generalized Born measurement on only the ancilla qubits, we rewrite the state $|\eta\rangle$ as follows

$$|\eta\rangle = \sqrt{p(0)}|0\rangle|\Phi_0\rangle + \sqrt{p(1)}|0\rangle|\Phi_1\rangle$$

where

$$|\Phi_0\rangle = \frac{1}{\sqrt{p(0)}}\sum_{x=0}^{2^n-1} a(x)|x\rangle\sqrt{1-v(x)} \quad ; \quad |\Phi_1\rangle = \frac{1}{\sqrt{p(1)}}\sum_{x=0}^{2^n-1} a(x)|x\rangle\sqrt{v(x)}$$

and

$$p(0) = \sum_{x=0}^{2^n-1} |a(x)|^2(1 - v(x)) \quad ; \quad p(1) = \sum_{x=0}^{2^n-1} |a(x)|^2 v(x)$$

Performing a generalized Born measurement measurement on only the ancilla qubit state, discussed in Ref. 4 yields

$$|\eta\rangle \quad \rightarrow \quad \text{Measurement} \quad \rightarrow \quad |a\rangle|\Phi_a\rangle \quad : \quad a = 0, 1$$

The probability of observing the ancilla qubit $|1\rangle$ is given by the following expectation value

$$\mu = p(1) = \langle\eta|\left(I_n \otimes |1\rangle\langle1|\right)|\eta\rangle = \langle\eta|1\rangle\langle1|\eta\rangle = \sum_{x=0}^{2^n-1} |a(x)|^2 v(x)$$

$$\Rightarrow \mu = E[v] \approx \int dx p(x)v(x) \qquad (9)$$

Hence, we see from Eq. (9) that any expectation value can be re-expressed as a quantum algorithm. The only constraint in finding the expectation value is that, since $0 \leq \mu \leq 1$, we must have that $v(x) \leq 1$ for all x.

Equation (9) is based on sampling a random outcome many times, and hence reproduces the classical Monte Carlo result. In observing the ancilla qubit we have two outcomes: $|1\rangle$ with probability μ and $|0\rangle$ with probability $1 - \mu$; this is a Bernoulli random variable with $\sigma_\mu^2 = \mu(1 - \mu)$. From the central limit theorem, performing the measurement N, the accuracy of the estimate is given by

$$\tilde{\mu} = \mu \pm \frac{\sigma_\mu}{\sqrt{N}} \quad \text{with } 66\% \text{ likelihood}$$

The accuracy and speed are the same as in the classical Monte Carlo calculation.

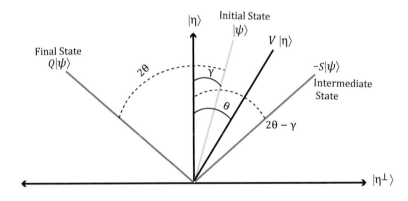

Fig. 1. The rotations of the initial qubit state $|\psi\rangle$ to the final state $Q|\psi\rangle$.

6. Algorithm for Quadratic Improvement

The application of quantum algorithms to option price has been discussed in Refs. 9–12. To get a quadratic improvement in the accuracy of the computed value of μ, we discuss the quantum algorithm for option price developed by in Ref. 9 — based on an oscillating phase.

The implementation of the oscillating phase approach is similar in spirit to the Grover algorithm, but the details are quite different. Note that the angle θ (defined below) is not known for option pricing, whereas in the

Grover algorithm, the angle θ is fixed by the total number of configurations of the degrees of freedom. What is similar to both approaches is that there is a rotation of an angle 2θ. For the option price, the rotation yields a final output qubits state for which the qubit that is most likely to be observed is the qubit that is closest in value to θ.

To perform the rotation, define the following unitary operator

$$V = I_{2^{n+1}} - 2I_{2^n} \otimes |1\rangle\langle 1| = V^\dagger \; ; \; V^2 = I_{2^{n+1}}$$

The state vector $V|\eta\rangle$ is a reflection of $|\eta\rangle$ about the axis defined by the state vector $|\psi\rangle$, and is shown in Fig. 1.

The angle between the state $|\eta\rangle$, defined in Eq. (7), and $V|\eta\rangle$ yields the angle θ, as shown in Fig. 1, and given by the following

$$\cos(\frac{2\pi\theta}{2}) \equiv \left(\langle\eta|\right)\left(V|\eta\rangle\right) = \langle\eta|V|\eta\rangle$$
$$= \langle\eta|\eta\rangle - 2\langle\eta|1\rangle\langle 1|\eta\rangle = 1 - 2\langle\eta|1\rangle\langle 1|\eta\rangle$$
$$= 1 - 2\sum_{x=0}^{2^n-1} |a(x)|^2 v(x) = 1 - 2\mu$$
$$\Rightarrow \mu = \frac{1}{2}(1 - \cos(\pi\theta)) \tag{10}$$

Once θ is evaluated, the price of the option μ can be obtained from it.

The rotation operator Q needs to be defined that will act on the input qubit $|\psi\rangle$ and rotate it in the plane spanned by $|\eta\rangle$ and $V|\eta\rangle$ through an angle 2θ to yield state vector $Q|\psi\rangle$. Following the derivation given in Ref. 9, define another operator

$$U = I_{2^{n+1}} - 2|\eta\rangle\langle\eta|$$

Note that

$$U|\eta\rangle = -|\eta\rangle \; ; \; U|\eta^\perp\rangle = |\eta^\perp\rangle \; : \; \langle\eta|\eta^\perp\rangle = 0$$

and hence $-U$ reflects about the axis defined by $|\eta\rangle$, leaving the state vector $|\eta\rangle$ unchanged. Define the operator

$$Z = I_{2^{n+1}} - 2|0\rangle^{\otimes(n+1)}\langle 0|^{\otimes(n+1)}$$

In terms of operator $\mathcal{F}$ introduced in Eq. (8), we have

$$U = \mathcal{F}Z\mathcal{F}^\dagger$$

since

$$U = \left[R(\mathcal{A} \otimes \mathbb{I}_2)\right]\left[I_{2^{n+1}} - 2|0\rangle^{\otimes(n+1)}\langle 0|^{\otimes(n+1)}\right]\left[(\mathbb{I}_2 \otimes \mathcal{A}^\dagger)R^\dagger\right]$$
$$= I_{2^{n+1}} - 2|\eta\rangle\langle\eta| \tag{11}$$

Define the diffusion operator Q, similar to the case of Grover's diffusion operator, that rotates the initial state vector $|\psi\rangle$ and is given by[9]

$$Q = U(VUV) \equiv US$$
$$= \left(I_{2^{n+1}} - 2|\eta\rangle\langle\eta|\right)V\left(I_{2^{n+1}} - 2|\eta\rangle\langle\eta|\right)V$$
$$= \left(I_{2^{n+1}} - 2|\eta\rangle\langle\eta|\right)\left(I_{2^{n+1}} - 2V|\eta\rangle\langle\eta|V\right) \tag{12}$$

since $V^2 = I_{2^{n+1}}$.

The action of $Q = US$, with $S = VUV$ and $U = \mathcal{F}Z\mathcal{F}^\dagger$, on an arbitrary state $|\psi\rangle$ in the span of $|\eta\rangle$ and $V|\eta\rangle$ is shown in Fig. 1. First, the action of $-S$ on $|\psi\rangle$ is to reflect along $V|\eta\rangle$, resulting in the intermediate $-S|\psi\rangle$. Then, $-U$ acts on $-S|\psi\rangle$ by reflecting along $|\eta\rangle$. The resultant state $Q|\psi\rangle$ has been rotated anticlockwise by an angle 2θ in the hyperplane of $|\eta\rangle$ and $V|\eta\rangle$.

From the expression above, it can be seen that operator Q rotates vectors in the two-dimensional plane defined by the vectors $|\eta\rangle$ and $V|\eta\rangle$, as shown in Fig. 1.

Note that

$$Q = UVUV = \mathcal{F}Z\mathcal{F}^\dagger \cdot V \cdot \mathcal{F}Z\mathcal{F}^\dagger \cdot V$$

The gates for operator Q are shown in Fig. 2.

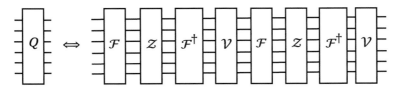

Fig. 2. Gates for operator Q.

7. Eigenvalues of Diffusion Operator Q

For the option pricing diffusion operator Q, the eigenvalues are evaluated, as this will allow us to isolate the angle θ that, via the relation $\cos \pi\theta = 1 - 2\mu$, will yield the option price.

Define the orthogonal vector $|\eta^{\perp}\rangle$ by

$$\langle\eta|\eta^{\perp}\rangle = 0$$

Using the identities

$$\langle\eta|V|\eta\rangle = 1 - 2\mu = \cos\pi\theta \ ; \ \langle\eta|\eta\rangle = \langle\eta|V^2|\eta\rangle = 1$$

yields

$$V|\eta\rangle = \cos(\pi\theta)|\eta\rangle + e^{iX}\sin(\pi\theta)|\eta^{\perp}\rangle$$

From the above equation it follows that, even though $|\eta\rangle$ and $V|\eta\rangle$ are not orthogonal, they are linearly independent.

Consider an arbitrary state vector $|\psi\rangle$ in the span of state vectors $|\eta\rangle$ and $V|\eta\rangle$; since these two vectors are linearly independent, we have

$$|\psi\rangle = a|\eta\rangle + bV|\eta\rangle$$

The action of Q on this state vector is given by

$$Q|\psi\rangle = aQ|\eta\rangle + bQV|\eta\rangle \tag{13}$$

Recall from Eq. (12)

$$Q = \left(I_{2^{n+1}} - 2|\eta\rangle\langle\eta|\right)\left(I_{2^{n+1}} - 2V|\eta\rangle\langle\eta|V\right)$$

Hence

$$Q|\eta\rangle = \left(-1 + 4\cos^2(\pi\theta)\right)|\eta\rangle - 2\cos(\pi\theta)V|\eta\rangle$$
$$QV|\eta\rangle = 2\cos(\pi\theta)|\eta\rangle - V|\eta\rangle \tag{14}$$

and from Eqs. (13) and (14) we obtain

$$Q|\psi\rangle = \left(a(-1 + \cos^2(\pi\theta)) + 2b\cos(\pi\theta)\right)|\eta\rangle$$
$$+ \left(-2a\cos(\pi\theta) - b\right)V|\eta\rangle \tag{15}$$

The action of the diffusion operator Q on the space spanned by $|\eta\rangle$ and $V|\eta\rangle$ is closed, and yields

$$Q|\psi\rangle = aQ|\eta\rangle + bQV|\eta\rangle$$
$$= \tilde{a}|\eta\rangle + \tilde{b}V|\eta\rangle \tag{16}$$

On the subspace spanned by $|\eta\rangle$ and $V|\eta\rangle$, Q is equal to the matrix M and yields the following linear transformation on the coefficients a, b

$$\begin{bmatrix} \tilde{a} \\ \tilde{b} \end{bmatrix} = M \begin{bmatrix} a \\ b \end{bmatrix} \tag{17}$$

The matrix M, from Eqs. (15) and (16), is given by

$$Q \equiv M = \begin{bmatrix} 4\cos^2(\pi\theta) - 1 & 2\cos(\pi\theta) \\ -2\cos(\pi\theta) & -1 \end{bmatrix} ; \quad \det M = 1 \tag{18}$$

The eigenvalues of M, given by λ, satisfy the equation

$$0 = \det(M - \lambda\mathbb{I}_2) = \det \begin{vmatrix} 4\cos^2(\pi\theta) - 1 - \lambda & 2\cos(\pi\theta) \\ -2\cos(\pi\theta) & -1 - \lambda \end{vmatrix} \tag{19}$$

and yields

$$(1 + \lambda)^2 - (1 + \lambda)\cos^2(\pi\theta) + \cos^2(\pi\theta) = 0$$

Hence the eigenvalues of Q are given by

$$\lambda_\pm = \frac{1}{2}\left(\cos^2(\pi\theta) - 2 \pm i\cos(\pi\theta)\sqrt{4 - \cos^2(\pi\theta)}\right) \tag{20}$$

Note that

$$\cos^2(\pi\theta) - 2 = 2\cos(2\pi\theta) ; \quad \cos(\pi\theta)\sqrt{4 - \cos^2(\pi\theta)} = 2\sin(2\pi\theta)$$

Hence, simplifying the eigenvalues given in Eq. (20) yields

$$\lambda_\pm = \cos(2\pi\theta) \pm i\sin(2\pi\theta) = \exp\{\pm i2\pi\theta\}$$

From Eq. (16), we have that the eigenvectors of Q are $|\psi_\pm\rangle$ with eigenvalues $\exp\{\pm i2\pi\theta\}$. The explicit representation of the eigenvectors is not required, but they are given for completeness. The eigenvectors of Q are given by

$$|\psi_\pm\rangle = a_\pm|\eta\rangle + b_\pm V|\eta\rangle$$

with

$$M\begin{bmatrix} a_\pm \\ b_\pm \end{bmatrix} = \begin{bmatrix} 4\alpha^2 - 1 & 2\alpha \\ -2\alpha & -1 \end{bmatrix}\begin{bmatrix} a_\pm \\ b_\pm \end{bmatrix} = e^{\pm i2\pi\theta}\begin{bmatrix} a_\pm \\ b_\pm \end{bmatrix} \qquad (21)$$

Using Eq. (21), the eigenvectors are the following

$$|\psi_+\rangle = \frac{1}{\sqrt{2}}\left[|\eta\rangle - e^{-i\pi\theta}V|\eta\rangle\right] \;\;;\;\; |\psi_-\rangle = \frac{1}{\sqrt{2}}\left[e^{i\pi\theta}|\eta\rangle + V|\eta\rangle\right] \qquad (22)$$

Note $|\psi_\pm\rangle$ are not normal or orthogonal since $|\eta\rangle$ and $V|\eta\rangle$ are not orthogonal. The amplitude amplification does not require that the eigenfunctions $|\psi_\pm\rangle$ be orthogonal, and hence the state vector $|\psi\rangle$ need not be expressed in an orthogonal basis.

8. Amplitude Amplification

The n-qubit $|\eta\rangle$ has the following eigenfunction expansion

$$|\eta\rangle = \beta_+|\psi_+\rangle + \beta_-|\psi_-\rangle$$

Consider a m-qubit state $|y\rangle$ and the $n+1$-qubit $|\eta\rangle$; the input tensor m-qubit $\otimes n + 1$-qubit state is given by

$$|y\rangle|\eta\rangle$$

Extending the definition of Q to Q_c such that it has the *conditional application*, depending on the m-qubit state $|y\rangle$ and is given by

$$Q_c = I_{2^m} \otimes Q \;\;\Rightarrow\;\; Q_c\Big(|y\rangle|\psi\rangle\Big) = |y\rangle Q^y|\psi\rangle$$

where

$$Q^y \equiv (Q)^y$$

The phase angle θ is evaluated in the following manner. Take a copy of input $|0\rangle^{\otimes(n+1)}$ and applying $\mathcal{F}$ yields

$$\mathcal{F}|0\rangle^{\otimes(n+1)} = |\eta\rangle$$

The m-qubit register is prepared in the uniform superposition state using the Hadamard gate and yields

$$H|0\rangle^{\otimes m}|\eta\rangle = \frac{1}{\sqrt{2^m}} \sum_{y=0}^{2^m-1} |y\rangle|\eta\rangle$$

Applying the controlled Q_c to the input state yields the following entangled state

$$|\Psi\rangle = Q_c \left[\frac{1}{\sqrt{2^m}} \sum_{y=0}^{2^m-1} |y\rangle|\eta\rangle \right] = \frac{1}{\sqrt{2^m}} \sum_{y=0}^{2^m-1} |y\rangle Q^y|\eta\rangle$$

$$= \frac{1}{\sqrt{2^m}} \sum_{y=0}^{2^m-1} |y\rangle Q^y \left(\beta_+|\psi_+\rangle + \beta_-|\psi_-\rangle \right)$$

$$\Rightarrow |\Psi\rangle = \frac{1}{\sqrt{2^m}} \sum_{y=0}^{2^m-1} |y\rangle \left(e^{2\pi i y\theta} \beta_+|\psi_+\rangle + e^{-2\pi i y\theta} \beta_-|\psi_-\rangle \right) \qquad (23)$$

The inverse Fourier transform on the $|y\rangle$ qubits is given by

$$|y\rangle = \frac{1}{\sqrt{2^m}} \sum_{x=0}^{2^m-1} e^{-2\pi i xy/2^m} |x\rangle$$

The full quantum circuit for the phase estimation is shown in Fig. 3.

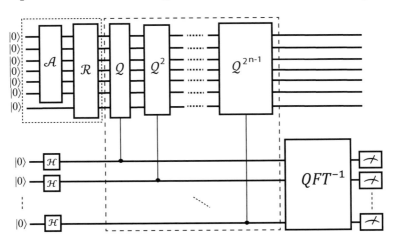

Fig. 3.   Quantum circuit for the estimation of the phase θ.

Hence, from Eq. (23)

$$|\Psi\rangle$$

$$= \frac{1}{\sqrt{2^m}} \sum_{y=0}^{2^m-1} \frac{1}{\sqrt{2^m}} \sum_{x=0}^{2^m-1} |x\rangle \left(e^{-2\pi i(x/2^m - \theta)y} \beta_+ |\psi_+\rangle + e^{-2\pi i(x/2^m + \theta)y} \beta_- |\psi_-\rangle \right)$$

$$\Rightarrow |\Psi\rangle = \sum_{x=0}^{2^m-1} |x\rangle \left(\beta_+(x) |\psi_+\rangle + \beta_-(x) |\psi_-\rangle \right) \tag{24}$$

where

$$\beta_\pm(x) = \frac{1}{2^m} \sum_{y=0}^{2^m-1} \exp\{-2\pi i(x/2^m \mp \theta)y\} \beta_\pm$$

Note from the above equation that x is given by

$$x = x_m 2^0 + x_{m-1} 2^1 + x_{m-2} 2^2 \cdots + x_i 2^i + \cdots + x_1 2^{m-1} \quad : \; 0 \le x \le 2^m - 1$$

Recall from Eq. (10) that the angle $\theta \in [0, 1]$ is defined by the payoff function, and is not an integer. Consider $\theta = \theta_m + \delta$, where θ_m is a binary decimal closest to θ. Then, for example

$$\beta_+(x) = \frac{1}{2^m} \sum_{y=0}^{2^m-1} \exp\{-2\pi i(x/2^m - \theta_m - \delta)y\} \beta_+$$

Note

$$\frac{1}{2^m} \sum_{y=0}^{2^m-1} \exp\{-2\pi i(x/2^m - \theta_m)y\} \simeq \delta_{x-2^m \theta_m} + O(\delta/2^m)$$

For $x = 2^m \theta_m$, we have

$$\beta_+(2^m \theta_m) = \frac{1}{2^m} \sum_{y=0}^{2^m-1} \exp\{2\pi i \delta y\} \beta_+ = \frac{1}{2^m} \frac{1 - e^{2\pi i \delta 2^m}}{1 - e^{2\pi i \delta}} \beta_+ \simeq \beta_+ + O(\delta/2^m)$$

Note that, from Eq. (24)

$$|\Psi\rangle = \sum_{x=0}^{2^m-1} \beta(x) |x\rangle \cdot \frac{1}{\beta} \left(\beta_+(x) |\psi_+\rangle + \beta_-(x) |\psi_-\rangle \right)$$

$$\Rightarrow |\Psi\rangle \equiv \sum_{x=0}^{2^m-1} \beta(x) |x\rangle |\Phi(x)\rangle \tag{25}$$

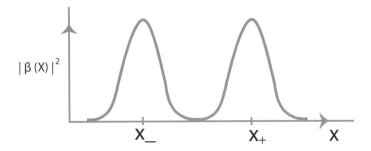

Fig. 4.   Probability distribution function for observing basis state $|x\rangle$.

On performing a generalized Born measurement of only the x-qubits in Eq. (25) we have

$$|\Psi\rangle \;\rightarrow\; \text{Measurement} \;\rightarrow\; |x\rangle|\Phi(x)\rangle \;:\; x = 0,1,\cdots,2^m - 1$$

The probability $\beta^2(x)$ to find the system in qubit basis state $|x\rangle$ is given by

$$|\beta(x)|^2 = \text{Tr}\Big(|x\rangle\langle x| \otimes I_{2^n}\Big)\Big(|\Psi\rangle\langle\Psi|\Big) = |\beta_+(x)|^2 + |\beta_-(x)|^2$$

The maximum likelihood of occurrence for a qubit x, as shown in Fig. 4, is given by

$$x_\pm/2^m = \pm\theta_m$$

Since the option price is the function $\cos(\pm 2\pi\theta_m)$, the value of either of the qubits $x_\pm$ that occur with maximum likelihood gives the option price.

It is shown in Ref. 9 that the estimate of the option price has the expected quadratic speeding up as in Eq. (5). An intuitive discussion of this result is given in Sec. 10.

9. Call Option

The price of a European call option is given by[13]

$$C(x_0, \tau) = \frac{e^{-r\tau}}{\sqrt{2\pi\tau\sigma^2}} \int_{-\infty}^{+\infty} dx\, e^{-\frac{1}{2\tau\sigma^2}(x - x_0 - \tau(r - \frac{1}{2}\sigma^2))^2}[e^x - K]_+$$

$$= e^{-r\tau} \int_{-\infty}^{+\infty} dx\, \frac{e^{-\frac{1}{2}x^2}}{\sqrt{2\pi}}\left[e^{\sqrt{\tau}\sigma x + x_0 + \tau(r - \frac{1}{2}\sigma^2)} - K\right]_+$$

$$\simeq e^{-r\tau} \int_{-x_{max}}^{+x_{max}} dx\, \frac{e^{-\frac{1}{2}x^2}}{\sqrt{2\pi}}\left[e^{\sqrt{\tau}\sigma x + x_0 + \tau(r - \frac{1}{2}\sigma^2)} - K\right]_+ \qquad (26)$$

In the approximation given in Eq. (26), one replaces $[-\infty, +\infty]$ by $[-x_{max}, +x_{max}]$; it is sufficient to take x_{max} to be 1. The probability density for the Gaussian random variable is given by

$$p(x) = \frac{1}{\sqrt{2\pi}} e^{-\frac{1}{2}x^2}$$

The random variable x is discretized into 2^n points with

$$x_j = -x_0 + j\Delta x \; ; \; \Delta x = 2x_{max}/(2^n - 1) \; ; \; j = 0, 1, \cdots, 2^n - 1$$

The discrete probabilities are given by the following

$$p(x) \rightarrow p_j = \frac{1}{c} p(x_j) \; ; \; c = \sum_{j=0}^{2^n-1} p(x_j)$$

The discretized probability distribution is shown in Fig. 5.

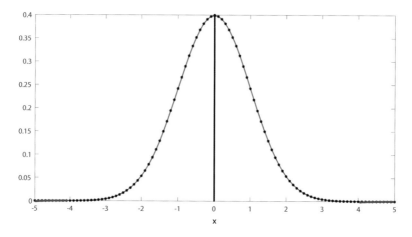

Fig. 5.   Discretized Gaussian probability distribution function.

The payoff function

$$v(x) = \left[e^{\sqrt{\tau}\sigma x + x_0 + \tau(r - \frac{1}{2}\sigma^2)} - K \right]_+$$

is discretized to

$$v_j = v(x_j)$$

A binary approximation $\tilde{v}_j$ is made of the discretized payoff function such that the error is given as

$$|v_j - \tilde{v}_j| < \frac{1}{2^n}$$

Hence, from Eq. (26), the approximate call option price is given by

$$C(x_0, \tau) \simeq e^{-r\tau} \sum_{j=0}^{2^n-1} p_j \tilde{v}_j \tag{27}$$

Equation (6) is specialized to

$$\mathcal{A}|0\rangle^{\otimes n} = \sum_{j=0}^{2^n-1} \sqrt{p_j} |j\rangle \tag{28}$$

The input qubit state, as in the general case, is prepared to yield

$$|\eta\rangle = \sum_{j=0}^{2^n-1} \sqrt{p_j}|j\rangle \left(\sqrt{1 - \tilde{v}_j}|0\rangle + \sqrt{\tilde{v}_j}|1\rangle \right) \quad : \quad \langle \eta|\eta \rangle = 1 \tag{29}$$

The option price, as in the general case, is given by

$$\mu = \langle \eta| \left(I_{2^n} \otimes |1\rangle\langle 1| \right)|\eta\rangle = \sum_{j=0}^{2^n-1} \tilde{v}_j p_j$$

and we have recovered the approximate call option price given in Eq. (27). The quadratic improvement discussed in Sec. 6 can be applied for pricing the call option.

10. Discussion

The Grover and option price algorithms are both based on amplitude amplification. There is however an interesting difference between the two. For Grover's case, one is seeking a marked state vector $|\xi\rangle$.[4] Hence, there is no need for a Fourier transform for isolating the marked state vector since doing the recursion enough times makes the coefficient of $|\xi\rangle$ in the superposed state vector almost equal to 1.

In contrast, the algorithm for finding the option price reduces the problem to finding an unknown angle θ. The Grover approach does not work since for Grover's algorithm, the angle θ is known — since it is determined by the number of qubits being used for the algorithm. The number of recursions required to find the marked state $|\xi\rangle$ is determined by θ.

For finding the option price, the Fourier transform is required to pick out the most likely state, which after sufficient iterations of the diffusion operator Q, is seen to yield the sought for angle.

Intuitively, for classical probabilistic models, the probability of success increases, roughly, by a constant on each iteration; by contrast, amplitude amplification roughly increases the amplitude of success by a constant on each iteration. Because amplitudes correspond to square roots of probabilities, it suffices to repeat the amplitude amplification process approximately $\sqrt{N}$ times to achieve the same accuracy, with high probability, as that of the classical case with N iterations.

Hence, the quantum algorithm discussed computes the option price more efficiently than the Monte Carlo simulations. For a simulation consisting of N trials, classical Monte Carlo yields an error that falls like $1/\sqrt{N}$, whereas in the case of a quantum algorithm there is a quadratic improvement and the error falls like $1/N$.

Acknowledgments

I thank Mahmudul Karim, Behzad Mansouri, L.C. Kwek and Frederick H. Willeboordse for many useful and fruitful discussions.

References

1. R. P. Feynman, *The Character of Physical Law* (Penguin Books, USA, 2007).
2. R. P. Feynman, *Lectures On Computation* (CRC Press, USA, 2018).
3. D. Deutsch, Quantum theory, the Church–Turing principle and the universal quantum computer, *Proc. Royal Society* (1985).
4. B. E. Baaquie and L. C. Kwek, *Quantum Computers: Theory and Algorithms* (Springer, 2022).
5. B. E. Baaquie, *Quantum Finance: Path Integrals and Hamiltonians for Options and Interest Rates* (Cambridge University Press, 2004).
6. B. E. Baaquie, *Interest Rates and Coupon Bonds in Quantum Finance* (Cambridge University Press, UK, 2010), 1st edition.
7. B. E. Baaquie, *Quantum Field Theory for Economics and Finance* (Cambridge University Press, 2018).
8. J. C. Hull, *Options, Futures, and Other Derivatives* (Prentice Hall, New Jersey, 2000), fourth edition.

9. P. Rebentrost, B. Gupt, and T. R. Bromley, Quantum computational finance: Monte Carlo pricing of financial derivatives, *Phys. Rev. A.* **98** (2018) 022321.

10. S. M. Román Orús and E. Lizaso, Quantum computing for finance: Overview and prospects, *Rev. Phys.* **4** (2019) 100028.

11. M. M. Gilles Brassard, Peter Hóyer and A. Tapp, Quantum amplitude amplification and estimation, arXiv:quant-ph/0005055v1 (2000).

12. N. T. Kazuya Kaneko, Koichi Miyamoto and K. Yoshinox, Quantum computing for finance: Overview and prospects, arXiv:1905.02666v4 (2020).

13. B. E. Baaquie, *Mathematical Methods and Quantum Mathematics for Economics and Finance* (Springer, Singapore, 2020).

Studies on Partition Temperature Model at LHC Energies

生
日
快
乐

A. H. Chan, C. W. Chong, C. H. Oh and Z. Ong

Department of Physics, National University of Singapore,
2 Science Drive 3, Singapore 117551, Singapore

The partition temperature model is a phenomenological model to interpret the multiparticle production of the inelastic hadron-hadron collisions. A brief review of the model allows one to gain insight into the motivations, derivations and some important features exhibited. A quick application is also done to verify the validity of the model at higher energy $\sqrt{s} = 0.9, 2.36, 7$ TeV.

1. Introduction

Multiparticle production in high energy collisions of particles arises mainly from strong interaction among their constituents at the peripheral regime where the coupling constant is relatively large. As is well-known, quantum chromodynamics provides a very good framework to explain and predict deep inelastic scattering reactions, but no satisfactory calculation technique is presently available to adequately account for the complete multiparticle phenomenology. There have been many phenomenological models devised to describe multiparticle production. Here we revisit the partition temperature model[1] which describes well the single particle angular distribution for the non-leading particles.

There have been numerous developments of the partition temperature model since the idea originated in Ref. 1 by T. T. Chou, C. N. Yang, and E. Yen and extended in Ref. 6 by Y. K. Lim, C. H. Oh and K. K. Phua. We will first briefly review this model as applied to the high energy hadron-hadron (hh) collision and explore the general behavior of the model (in Sec. 2). We

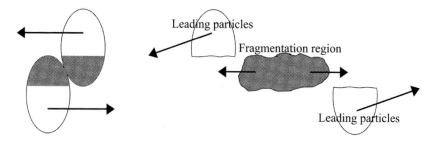

Fig. 1.   Schematic diagrams for multiparticle production. The overlapping part between the particles (or central region) coalesces and exchanges longitudinal and transverse momenta, then fragments into many fast particles, while the rest of the particles play the role of leading particles, continuing their trajectory without being interrupted.[2]

then apply it to the recent Large Hadron Collider (LHC) data (in Sec. 3) and discuss the result of the fitting.

2. Partition Temperature Model — hh Collision

2.1. *Pseudorapidity distribution of nonleading particles*

As pointed out by T. T. Chou and C. N. Yang, at each impact parameter b, the total charged multiplicity $n = n_F + n_B$ exhibits the Koba–Nielsen–Olesen scaling (KNO scaling),[3] where the charge asymmetry that corresponds to the difference between the forward multiplicity and backward multiplicity $Z = n_F - n_B$ appears to follow the binomial distribution,[4] as observed through the data from proton-antiproton collisions ($p\bar{p}$) collisions at $\sqrt{s} = 540$ GeV from the UA5 experiment.

The forward-backward multiplicity distribution forms an ellipse in the two-dimensional $\frac{n_F}{\bar{n}}$ vs $\frac{n_B}{\bar{n}}$ plane as illustrated in the diagram below, where $\bar{n}$ denotes the average charged multiplicity. The ellipse can be viewed as the superposition of small circles, each of which represents a stochastic fluctuation. At higher energy, as more particles would be produced per event, $\bar{n} \to \infty$, it is claimed that this ellipse will coalesce to a straight line and thus we have $n_F \approx n_B$, which means that only stochastic fluctuations are present. Therefore, the non-stochastic (approximate KNO scaling) fluctuation of the total charge multiplicity can be considered as the incoherent superposition of collisions at different impact parameters.[1,2]

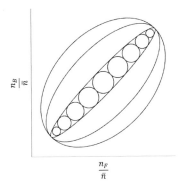

Fig. 2. Schematic diagram of scaled forward-backward multiplicity distribution $P(\frac{n_F}{\bar{n}}, \frac{n_B}{\bar{n}})$, the contour lines shrink to the form of a thin line at very high energies.[2]

Such stochastic behavior then motivates the stochastic assumption on the energy partition on each side of hemisphere at fixed KNO variable $z = \frac{n}{\bar{n}}$, subject to some constraints:

- The conservation of energy $\delta(\sum_i E_i - h E_0)$
- The transverse momentum cut-off factor $g(p_{\perp,i})$
- The Lorentz Invariant Phase Space (LIPS) $\prod_i \frac{d^3 p_i}{E_i}$

Here $E_0 = \frac{\sqrt{s}}{2}$ denotes the incoming energy of a colliding particle, h denotes the fraction of the incoming energy E_0 that participated in the multiparticle production in the central region, and so $E_0(1 - h)$ denotes the total energy carried away by the leading particles. Note that i ranges over all the particles produced on one side of the hemisphere minus the leading particles.

Putting all these components together, the probability distribution of non-leading particles on each hemisphere in the centre of momentum (CM) frame is:

$$dn = \delta(\sum_i E_i - h E_0) g(p_{\perp,i}) \prod_i \frac{d^3 p_i}{E_i} \tag{1}$$

However, this expression does not seem useful enough to understand how the energy is partitioned since h cannot be determined accurately, as it is ambiguous to differentiate between the particles, whether they are the leading particles, or the particles produced in the central region, especially for larger values of pseudorapidity η. Also, since there is energy exchange during collisions, by analogy with statistical mechanics, the expression which appears to be a microcanonical distribution can be modified into a canonical distribution by ignoring the correlation effects. As a result, the stochastic distribution for a single charged particle produced in each hemisphere in the CM frame can be written as follows:

$$dn = Ke^{-\frac{E}{T_p}}g(p_\perp)\frac{d^3p}{E} \tag{2}$$

where T_p is a parameter known as partition temperature that controls the energy partition in the hemisphere, K is the normalization constant that can be adjusted, and $E^2 = p^2 + m^2$. The presence of K is due to the difference between the normalization constants of the microcanonical distribution and that of the canonical distribution.

An additional simplification can be done to make the expression more useful in the calculation. By expressing the integral in spherical coordinates and with the relationship between polar angle and pseudorapidity $\cosh \eta = \frac{1}{\sin \theta}$, we then obtain the single charged particle angular distribution:

$$\frac{dn}{d\eta} = K\frac{2\pi}{\cosh^2 \eta}\int_0^{p_{max}} e^{-\frac{E}{T_p}}g(p_\perp)\frac{p^2dp}{E} \tag{3}$$

where the maximum momentum p_{max} is simply the incoming energy that participated in the multiparticle production hE_0.[1]

2.2. Choice of the transverse momentum cut-off factor

To perform the integration, it is necessary to have an explicit form of the transverse momentum cut-off factor, which should vanish at large transverse momentum. Since we assume that at high energy the energy distribution in each hemisphere is stochastic, the pseudorapidity distribution should be an even function of η. It is also worth noting that the transverse momentum cut-off factor must be dimensionless.

It is natural to choose $g(p_\perp) = e^{-\alpha p_\perp}$ or $g(p_\perp) = e^{-\alpha' p_\perp^2}$ which satisfy the conditions mentioned above, where α and α' are parameters to be fixed. In Ref. 1, the authors chose $g(p_\perp) = e^{-\alpha p_\perp} = e^{-\frac{\alpha p}{\cosh \eta}}$, where $\alpha = \frac{2}{p_\perp}$, taking into account the transverse momentum distribution. We will then consider such choice for the rest of the paper.

2.3. *Application to the experimental result*

Having explicitly described all components, one can now evaluate the pseu-dorapidity distribution $\frac{dn}{d\eta}$ for any given η and fit the resulting curve to the experimental data points. For example, here in Fig. 3, the data are provided by the UA5 experiment with $p\bar{p}$ collisions and the curves are fitted by ad-justing the normalization constant K, with a value of T_p determined for each multiplicity n. The multiplicity is used here to distinguish the collision in-stead of the impact parameter since the multiplicity is closely related to the impact parameter (as an implicit assumption), and the impact parameter is not directly accessible in the experiment.[5] The graph shows that the curve agrees with the experimental data, especially for small η ($0 \leq \eta \leq 4$), where the data are more accurate.[2] As a result, the partition temperature model is substantiated by the experimental result and can provide a good description for collisions with multiparticle production.

When discussing the canonical ensemble in statistical mechanics, the temperature manifests as the Lagrange multiplier for the constraint of the average energy, and consequently, it implies thermal equilibrium. Here, as the collisions that we are interested in are small scaled and occur in a short time, it is clear that the thermal equilibrium cannot be achieved. Indeed, the partition temperature T_p does not imply thermal equilibrium in the model, it is simply a parameter that controls the way energy is partitioned on each side of the collisions.

2.4. *Properties of the pseudorapidity distribution function*

By assuming that the majority of the particles produced in the fragmentation region are light charged mesons, one can make the approximation $E \approx p$. Also due to the fact that we focus mainly on small η situation where the data is more accurate, and as an experimental fact $\alpha \gg \frac{1}{T_p}$, the integration in the massless pseudorapidity distribution expression can then be performed and we obtain[1]

$$\frac{dn}{d\eta} = K\frac{2\pi}{\left(\frac{\cosh\eta}{T_p} + \alpha\right)^2} \tag{4}$$

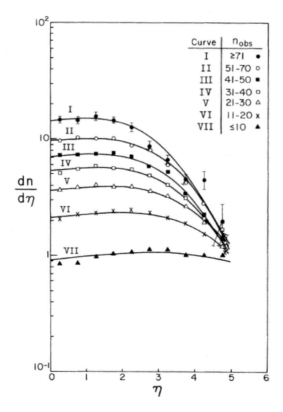

Fig. 3. Plot of $\frac{dn}{d\eta}$ as a function of η for different multiplicities. Data points are taken from the UA5 experiment $\sqrt{s} = 540$ GeV while the curve is determined by curve fitting of pseudorapidity distribution shown previously.[1]

We see that such function is an even function of η, and a monotonically decreasing function for $\eta > 0$ with maximum value occuring at $\eta = 0$. We also observe that it has a sharp cutoff at large η due to the nature of $\frac{1}{\cosh^2 \eta}$. Graphically, unlike the distribution shown from the experimental result, the distribution illustrates the absence of the valley near $\eta = 0$, meaning that the presence of the mass term contributes to the plateau structure. Hence, in the general expression for pseudorapidity distribution, we expect a deeper valley when more massive particles are produced at higher energy.

2.5. *Fraction of incoming energy in multiparticle production*

One of our main interests in multiparticle production collisions is to determine the proportion of the total energy carried away by the leading particles $(1 - h)$ and the proportion of the total energy participating in the collision process h. In some literature contexts, such as cosmic-ray physics, the factor h is known as inelasticity.[2] As mentioned, the value of inelasticity is ambiguous, and yet the success of the partition temperature model allows us to eliminate this ambiguity. Once the partition temperature T_p is determined by fitting the curve, we have a well determined pseudorapidity distribution, and so the fraction of the incoming energy that participated in the multiparticle production h can be computed as

$$hE_0 = \frac{3}{2} \int_0^\infty d\eta \int_0^{p_{max}} K \frac{2\pi}{\cosh^2 \eta} e^{-\frac{\sqrt{p^2+m^2}}{T_p}} g(p_\perp) p^2 dp \qquad (5)$$

Note that we need to add a numerical factor $\frac{3}{2}$ since the expression holds for all the charged particles, and in general, there are positively charged, negatively charged and neutral particles produced from the collisions.

Besides, a similar calculation can be performed to find the total multiplicity in the collision process:

$$n = \frac{3}{2} \int_0^\infty d\eta \int_0^{p_{max}} K \frac{2\pi}{\cosh^2 \eta} e^{-\frac{\sqrt{p^2+m^2}}{T_p}} g(p_\perp) \frac{p^2 dp}{\sqrt{p^2 + m^2}} \qquad (6)$$

2.6. *Global partition temperature*

As suggested by Y. K. Lim, C. H. Oh, and K.K. Phua,[6] instead of considering a single partition temperature for each impact parameter, it is possible to fit the overall pseudorapidity distribution that sums all the pseudorapidity distribution for different impact parameters (or multiplicity) using a global partition temperature. They used pseudorapidity data from CERN collider and CERN ISR for the fitting at different energy levels ranging from 30.4 GeV to 900 GeV. In general, the fitting results look good at $\eta < 4$, and there are significant errors only at the tail of the distribution with large η for some energy levels, where the data has large uncertainty.

Also, inspired by the negative binomial distribution of the charged multiplicity, they found that the mean partition temperature should be proportional to the average energy of the particle produced in the hadron–hadron collisions, more precisely this means $\bar{T}_p \propto \frac{hE_0}{\bar{n}}$, where $\bar{n}$ denotes the average multiplicity.[6]

2.7. *A remark on the pp and $p\bar{p}$ collisions*

Fundamentally, the two hadron–hadron collisions should be different from each other due to the difference in the conserved quantities, namely quark number and electric charge. We should then expect to have different pseudorapidity distributions for these collisions. Surprisingly, the comparison between the experimental data of pp collisions collected in 2009 at the LHC[7] and the experimental result of $p\bar{p}$ collisions reported by the UA5 experiment at the same energy level 0.9 TeV show that there is no significant difference between their distributions. This means that the partition temperature model should work for all types of hadron–hadron collisions, regardless of their fundamental difference, as the model is derived simply from a statistical point of view.[7]

3. Application of the Model
3.1. *Experimental data at higher energy*

We can also apply the partition temperature model to the data at higher energy, collected at the LHC in 2009 and 2010,[7,8] where the pseudorapidity and the transverse momentum distributions are reported for pp collisions at the energy level $\sqrt{s} = 0.9, 2.36, 7$ TeV. Assuming that the particles produced are all pions, and with the average transverse momentum determined from the corresponding distribution, we may then use the curve fitting method to fit the pseudorapidity distribution to the data points given that the parameter K and T_p are determined for each energy level.

3.2. *Results*

For comparison, the curve fitting is performed with $g(p_\perp)$ as (i) $e^{-\alpha p_\perp}$ and (ii) $e^{-\alpha' p_\perp^2}$, where α has the same definition as proposed by the authors of the partition temperature model,[1] $\alpha = \frac{2}{\bar{p}_\perp}$, and we use $\alpha' = \frac{1}{\bar{p}_\perp}$. The fitting results are shown in the graph below.

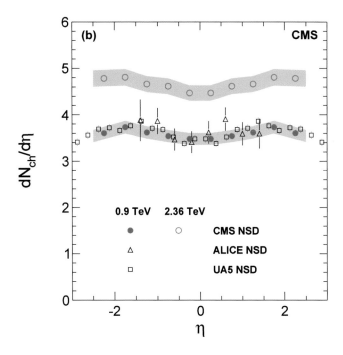

Fig. 4.   Reconstructed $\frac{dn}{d\eta}$ distributions averaged over the cluster counting, tracklet and tracking methods (circles), compared to data from the UA5 (open squares) and from the ALICE (open triangles) experiments at 0.9 TeV.[7]

To assess the quality of the distribution, we may borrow the idea of the chi-squared test to check how well the distribution fits the data. More explicitly, we compute $\chi^2 = \sum_\eta \frac{(O_\eta - E_\eta)^2}{E_\eta}$ where O_η is the value computed by the determined pseudorapidity function and E_η is the value of the experimental pseudorapidity data points, for each given η. If the distribution fits well with the data points, a small value of χ^2 will be expected since there is less error from the experimental data points.

With K and T_p determined from the above curve, we can calculate the inelasticity h and total multiplicities n in each hemisphere for each energy level. The results are presented in the tables below.

3.3. *Discussions*

We see that both results have a fairly small value of χ^2 at different energy levels, and both results show a similar trend for the partition temperature T_p

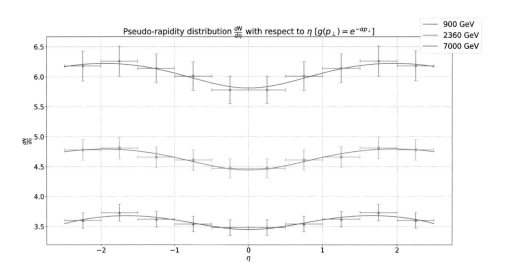

Fig. 5.   Plot of $\frac{dn}{d\eta}$ as a function of η for $\sqrt{s} = 0.9, 2.36, 7$ TeV using $g(p_\perp) = e^{-\alpha p_\perp}$.

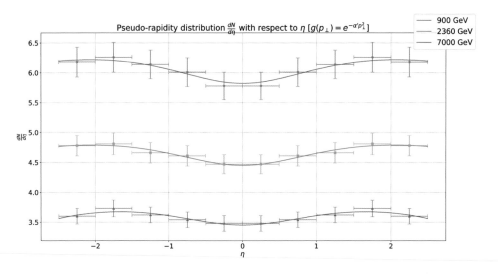

Fig. 6.   Plot of $\frac{dn}{d\eta}$ as a function of η for $\sqrt{s} = 0.9, 2.36, 7$ TeV using $g(p_\perp) = e^{-\alpha' p_\perp^2}$.

and the normalization constant K, yet there is a large difference in the value between the two choices. The difference in K is clear since the pseudorapidity expression, which is normalized to the number of particles detected, is different. Nevertheless, the total multiplicities seem to be consistent for both choices.

Table 1.   The determined parameters from the curve fitting and the associated results, using $g(p_\perp) = e^{-\alpha p_\perp}$.

Energy level $\sqrt{s}$	0.9 TeV	2.36 TeV	7 TeV
Partition temperature T_p (GeV)	25.01	60.94	75.03
Normalization constant K (GeV)$^{-2}$	12.04	12.82	13.93
Inelasticity h	0.17	0.19	0.10
Total multiplicities n	25.4	38.1	50.5
Chi-squared value χ^2	0.0024	0.0019	0.0023

Table 2.   The determined parameters from the curve fitting and the associated results, using $g(p_\perp) = e^{-\alpha' p_\perp^2}$.

Energy level $\sqrt{s}$	0.9 TeV	2.36 TeV	7 TeV
Partition temperature T_p (GeV)	27.32	88.30	115.36
Normalization constant K (GeV)$^{-2}$	5.91	6.30	6.85
Inelasticity h	0.18	0.28	0.16
Total multiplicities n	25.7	40.5	54.2
Chi-squared value χ^2	0.0026	0.0017	0.0031

On the other hand, the partition temperature and the resulting inelasticity seem to be vastly different from each other. This could possibly be due to the large uncertainty in the data points given, and only very few data points are available to fit the curve. The fitting result may then have a lot of uncertainty since it does not even include some larger η that lead to the sharp cut-off occurring in the distribution, due to the limitation of the detector.

Therefore, although the fitting result seems to be good graphically, to have an accurate result of the partition temperature and other variables, it is necessary to include more data points at different η with small error bars.

4. Remarks

We have seen how the partition temperature model is derived from the assumption inspired by the stochastic distribution of charge asymmetry in high-energy collisions. The concept of the partition temperature emerges by drawing the analogy from the study of the ensemble in statistical mechanics. It arises simply from the entropy maximization with the constraint of energy conservation while allowing energy exchange. By determining the parameters appearing in the pseudorapidity distribution function through curve fitting, the fraction of the incoming energy participating in the multiparticle production, or simply inelasticity h can be found. The success of the partition temperature model eliminates the ambiguity of treating fast particles as leading particles or particles produced in the central region.

The fitting results at higher energy level show again that the pseudorapidity distribution can be fitted well, however, we noticed that there are many differences in the values of parameters determined by the model using different choices of transverse momentum cut-off factors. This could be mainly due to the reason that the experimental data have too few data points with large error bars, which could introduce many uncertainties in determining this distribution. Yet, we find that there are no significant discrepancies in the application of the partition temperature model.

In the future, as more data become available, in order to ensure that this phenomenological model provides a good description for the experimental data, application to higher energy data is necessary. Perhaps the advanced experimental setup and reconstruction algorithm of the experimental detections will be able to provide more data with less uncertainty so that the validity of the partition temperature model can be verified. Nevertheless, even if a significant discrepancy is found in the future, we have seen that it works quite well for the current LHC data. Therefore, it is likely that only a small modification is needed, the concept of partition temperature should remain.

References

1. T. T. Chou, C. N. Yang, and E. Yen, Single-particle momentum distribution at high energies and concept of partition temperature, *Phys. Rev. Lett.* **54**(6), 510–513 (1985).
2. T. T. Chou and C. N. Yang, Geometrical model of multiparticle production in hadron-hadron collisions, *Phys. Rev. D* **32**(7), 1692–1698 (1985).
3. T. T. Chou, Remarks on multiplicity fluctuations and KNO scaling in $\bar{p}p$ collider experiments, *Phys. Lett.* **116**(4), 4 (1982).
4. T. Chou and C. N. Yang, Binomial distribution for the charge asymmetry parameter, *Phys. Lett. B* **135**(1-3), 175–178 (1984).

5. T. S. Li and K. Young, Single-particle distribution in hadron-nucleus scattering and the partition-temperature model, *Phys. Rev. D* **34**(1), 142–147 (1986).
6. Y. K. Lim, C. H. Oh and K. K. Phua, Charged-particle multiplicity distribution and partition temperature, *Phys. Rev. D* **38**(3), 1004–1007 (1988).
7. CMS Collaboration, Transverse-momentum and pseudorapidity distributions of charged hadrons in pp collisions at $\sqrt{s} = 0.9$ and 2.36 TeV, *J. High Energy Phys.* **2010**(2), 41 (2010).
8. CMS Collaboration, Transverse-momentum and pseudorapidity distributions of charged hadrons in pp collisions at $\sqrt{s} = 7$ TeV, *Phys. Rev. Lett.* **105**(2), 022002 (2010).

Phase Transition in Complex Systems: A Scientific Journey

生
日
快
乐

Lock Yue Chew

Division of Physics and Applied Physics,
School of Physical and Mathematical Sciences, SPMS-PAP-04-04,
Nanyang Technological University, 21 Nanyang Link, Singapore 637371
lockyue@ntu.edu.sg

This paper reviews my research on the phase transition of three types of complex systems that were positively impacted by Prof. Kok Khoo Phua and World Scientific Publishing through their support and engagement. The three phase transition scenarios are: (1) protein phase transition, (2) self-organized criticality (SOC) in fractal lattices, and (3) adaptive SOC in a coupled social-ecological system. The review highlights the subtle connections between phase transition in the three complex systems and the theory of spin glass and SOC, and through these connections illustrate a potential unification of concepts that these two major theories of complex systems could bring to the physics of phase transition in complexity science.

1. Introduction

In celebrating Prof. Kok Khoo Phua's 80th birthday, I would like to express my deep gratitude to him and World Scientific Publishing for their positive impact on my scientific career by presenting this research article. My relationship with Prof. Phua began more than 15 years ago when he was the director of NTU's Institute of Advanced Studies (IAS). Through Prof. Phua's engagement of Prof. Kerson Huang, Emeritus professor of MIT, as visiting professor in NTU, I started a collaboration with Prof. Huang that launched my research in the area of phase transitions in complex systems. It was in 2008 that IAS supported us for three years with a research grant known as the "Protein Folding Research Project". The funding enabled us to form a research team constituting: Prof. Kerson Huang, myself, Dr. Jinzhi Lei and Dr.

Weitao Sun (visiting scientists from Tsinghua University), Dr. Jerome Benoit and Dr. Xiaohui Qu (research fellows), Dr. Hon-Wai Leong and Dr. Lina Zhao (PhD students), Mr. Yee Woon Lim (Master student), Dr. Boon Chong Goh and Mr. Zhong Yi Chia (undergraduate students). The research collaboration lasted for about 5 years and was a productive one, with the eventual publications of more than 10 research papers. The research experience had benefited all of us in different ways. For me, it had brought a new perspective to the physical modeling of many-body system (such as proteins) through the approach of statistical mechanics.

In addition, I have formed linkages with World Scientific Publishing by serving as editor for its journal *Fractals: Complex Geometry, Patterns, and Scaling in Nature and Society* since 2013. Subsequently, I worked on a book project with World Scientific and successfully published a textbook: *Lecture Notes on Mechanics: Intermediate Level* in 2020. Recently, I have participated as guest editor for its journal *Advances in Complex Systems (ACS)* on a Special Issue: *Smart Cities of the Future*. This is an interesting project for which I hope to fruitfully contribute.

A way to acknowledge Prof. Phua and World Scientific Publishing for positively influencing my academic development is to present my research work that closely connects with this influence. In this context, I will like to detail my research relating to the area of "protein", "fractals" and "social physics". The principal focus of my presentation will be on the topic of phase transition within each of these subject areas. For this purpose, I will review my research and its outcome in protein phase transition, self-organized criticality in fractal lattices, and critical transitions in coupled social-ecological systems in the next three sections of this article. I will then conclude the article by outlining the potential unification of concepts that the theory of spin glass and SOC could bring to the physics of phase transition in complex systems.

2. Protein Phase Transition

Proteins exhibit the features of complex systems. Whether it is with regards to its conformational dynamics or its folding from denatured state to folding state, a theory of complex systems based on spin glass provides a good description of the processes.[1] In fact, the contribution of spin glass to the understanding of complex systems was duly recognized when part of the 2021 Nobel Prize in Physics was awarded to Giorgio Parisi. Spin glass is a good theoretical construct of a complex system from a physics perspective. It is a model in condensed matter physics that is related to disordered materials, and is characterized by quenched disorder and frustration. As a result, it possesses a complicated state space structure with many metastable or stable spin configurations. This implies the presence of many hills and valleys if we were to take the point of view of a free energy landscape. Such a landscape is rugged and in fact corresponds to that of a protein when we account for the thermodynamics and kinetics as a protein fold. This is the so-called "folding-funnel" concept of protein folding.[2]

When the protein is at the top of the energy funnel, it is at a denatured state (or in a disorder phase) which is not a stable configuration. The protein thus proceeds to fold and moves down the landscape through kinetic pathways. But before long, the protein becomes stuck in metastable states which exhibit the glassy phase of frozen-in disorder.[3] Here, the dynamics are slow and is known biologically as the *molten globule* phase of proteins. Dr. Jinzhi Lei and Prof. Kerson Huang had studied the collapse of the protein from denatured state to a molten globule by using the Conditioned Self-Avoiding Walk (CSAW) model[4] with hydrophobic effect and hydrogen bonding but without considering the electrostatic and van der Waals interactions. They deduced scaling law of the form $R_g \sim N^\nu$ where R_g is the radius of gyration and N is the number of amino acid residues. In their study, they had considered a pre-globule phase just before the protein adopts the phase of molten globule. They found $\nu = 3/5, 3/7, 2/5$ at the unfolded stage, at the pre-globule and the molten globule, respectively.[5] The phase of pre-globule goes over to the molten globule phase as a first-order phase transition with a sudden acceleration of hydrogen bonding. The molten globule plays an intermediate role in the folding process. It is a phase when the protein is partially folded between the unfolded states and the native state, before the protein eventually folds into its native state.

To study the statistical mechanics of protein folding and to explore protein phase transition, Dr. Hon-Wai Leong, myself and Prof. Kerson Huang formulated a Hamiltonian with the torsion angles of the protein chain as generalized coordinates (q), and hydrogen bonding as interaction energies ($U(q)$), as follows:

$$H = \frac{1}{2}p^T M^{-1} p + U(q),$$

where p is the generalized momenta.[6] The Hamiltonian formalism allows us to study the dynamics of protein folding through the normal modes of the protein secondary structures. Here, we explored protein phase transition from the native structure of alpha-helix to beta-sheet, and from beta-sheet to random coil, at two different critical temperatures. A more rigorous analysis was also performed by constructing a grand partition function of the protein, which enabled us to deduce that the two transitions are first-order phase transitions.[7]

These protein phase transition results are verified by numerical computation using CSAW.[4] CSAW is a stochastic approach in protein folding where the protein's atoms and molecules are assumed to undergo Brownian motion via self-avoiding walk. Because the protein interacts with itself and also the surrounding water, the self-avoiding walk needs to be conditioned by the different types of interactions encountered by the protein. As a theoretical laboratory, we could successively add each of these interactions into CSAW to study their effects. Note that CSAW is run through a Monte Carlo procedure using the Metropolis algorithm.

Recently, there has been a revolution in protein folding prediction where a software known as AlphaFold dominated the contest known as the Critical Assessment of Protein Structure Prediction (or CASP) by predicting the 3D shapes of protein from their amino-acid sequence that is on par with the experimental structures.[8] AlphaFold uses the machine learning technique of deep learning and attention-based transformation modules to exceed the performance of all existing approaches. Currently, it has made possible the prediction of protein structures of nearly every protein made by humans as well as those of mice and *Escherichia coli*. However, the software has the limitation of not being able to make predictions when the effect of mutations in the proteins is to be considered or when the proteins were to adopt different structures in different conformations. To overcome this limitation, the incorporation of protein physics into the software is required such that it can predict the dynamics of protein folding from the unfolded state to the folded state. This may be achieved by developing a physics-based machine learning approach where a combination of prior knowledge from constraints imposed by the physical laws together with data-driven learning is used to expand the strength of the techniques towards dynamical modeling. Such an advance is especially important in predicting protein misfolding which is the cause of "mad-cow" disease, Creutzfeld–Jacob disease and possibly Alzheimer's disease.

3. Self-Organized Criticality in Fractal Lattices

In addition to spin glasses, another candidate of the theory of complex systems is self-organized criticality (SOC). The idea of SOC was advanced by Per Bak through the paradigm of a sandpile metaphor.[9] In this metaphor, grains of sand are dropped gradually on a tabletop and after some time a pile is formed. At this stage, dropping a small amount of sand can trigger an avalanche, and the state of the system is said to be *critical*. At this critical state, the sandpile develops a threshold slope where an avalanche will decrease it locally while the dripping sand will increase it again. This is a nonequlibrium steady state where the influx of matter (sand) and energy equals the outflux of matter and energy. Through the driven sand, the system (sandpile) *self-organized* into a steady state at which the slope of the pile fluctuates about the threshold slope.[10] It is interesting that at this state, the system displays the features of critical transition. If we were to measure the size of the avalanches and evaluate its statistical behavior, a power-law distribution will be revealed. This is akin to a second-order phase transition in an Ising model where the cluster size distribution of the spin would also exhibit a power-law, although the critical point is reached through parameter (e.g. temperature) tuning. The underlying mechanism that gives rise to the power-law behavior in the Ising model is that of a balance between order and disorder. In the case of SOC, there is no tuning of parameters, and the mechanism that generates its behavior is the balance between an active phase (with activity) and an absorbing phase (the activity dies out).[11]

SOC is a compelling theory of complex systems because it seems to explain the ubiquitous empirical observations of power-law behavior in nature, such as in earthquake magnitude, cotton price, ranking of cities or words, brain signals, biological extinction, pulsar glitches, etc. It was formulated to understand the common observation of $1/f$ noise in the natural world. To achieve this understanding, Per Bak, Chao Tang and Kurt Wiesenfeld developed a numerical model of the sandpile known as the BTW model.[12] BTW model is a deterministic model with updated rules on the accumulation of sands on the lattice as grains of sand are dropped on it, with a threshold that signals the event of toppling which initiates avalanches. By observing the distribution of cluster size or lifetime of the avalanches from the numerical simulation, one would then find the occurrence of power-law behavior.

Like critical behavior in second-order phase transition, SOC is associated with the notion of *universality class*. A key feature of systems in the same universality class is that they share the same value for their critical exponents which obey a certain set of scaling laws. For example, the liquid–gas phase transition is generally placed in the same universality class as that of the Ising magnets and their critical exponents have the same value. In the same vein, the Manna model is considered to be a universality class in SOC. It is in fact the largest universality class with models that are known to display robust scaling behavior.[13]

We had used the Abelian Manna Model (AMM), which is one of the models in the Manna model, to explore SOC in our research. AMM is a stochastic SOC model and it possesses the aforementioned robustness property of the Manna model. Moreover, instead of using regular lattices as sites for accumulating the sand grains, we used fractal lattices to investigate the effect of lattice dimension on the scaling behavior of SOC.[14] Our aim is to uncover the systematic dependence of the critical exponents on the dimension and structure of the lattice. For this goal, we examined the continued validity of the scaling law

$$D(2 - \tau) = 2, \tag{1}$$

when the dimension of the lattice is fractal. Equation (1) is known to hold for the hypercubic lattice independent of its dimension. Note that D and τ are critical exponents of the probability density $P(s)$ for an avalanche of size s to occur:

$$P(s) = \alpha s^{-\tau} G\left(\frac{s}{\beta L^D}\right), \tag{2}$$

where L is the linear system size, α and β being nonuniversal metric factors, and $G()$ is a universal function representing finite-size scaling.

Interestingly, we found that the scaling law given by Eq. (1) is violated for the two fractal lattices: Sierpinski arrowhead and crab, even though these two lattices have the same lattice dimension of 1.58. The reason for the violation is that the critical exponents of fractal lattice depend not only on the dimension but also on its microscopic details. The consequence is that the critical exponents D and τ of the two fractal lattices can both be different. Remarkably, the scaling law is fulfilled if we were to generalize it to the following form:

$$D(2 - \tau) = d_w, \tag{3}$$

where d_w is the fractal dimension of random walk on the lattice. Equation (3) turns out to be true for both the hypercubic lattice and also the fractal lattice. For hypercubic lattice, $d_w = 2$, while $d_w = 2.322$ for the Sierpinski arrowhead lattice and $d_w = 2.578$ for the crab lattice.

The outcome of our studies on SOC in fractal lattices have brought to our attention the following question: could critical exponents of regular lattice depend not only on its dimension but also on its microscopic structure? We realized that this question had not been answered in the literature. As such, we conducted detailed numerical studies on eight two-dimensional and four one-dimensional lattices with different microscopic structure to find out. The eight 2D lattices we explored are: square, jagged, Archimedes, non-crossing diagonal square, triangular, kagome, honeycomb and mitsubishi, while the four 1D lattices are: simple chain, rope ladder, next nearest neighbor chain and futatsubishi. Our results show that both Eqs. (1) and (3) are valid for all of these regular lattices since $d_w = 2$ for all of them.[15]

Currently, the theory of SOC is being actively pursued in understanding the complex processes in the brain. There is a hypothesis that the brain may be poised in a self-organized critical state and this hypothesis has been supported by considerable amount of experimental evidence in the mammalian cortex. The basis of this conjecture is that close to a critical point, the brain is sensitive towards perturbation such that the amount of information transferred within it is maximal.[16]

4. Critical Transition in Coupled Social-Ecological System

It is interesting that SOC also occurs in social system. We had uncovered its presence in the Balinese subak, which is a self-organized society without centralized governance. A subak is an agrarian society of farmers who exploit water, which is a common pool resource, for irrigation and rice production activities. For more than 2000 years, this society adopted a philosophy that unites the realms of the spirit, the human, and nature.[17] Because of which, a strong connection has been established between the social and the ecological world. We studied this coupled social-ecological system to understand how the farmers manage the rice terraces cooperatively as they face irrigation water shortages and harvest loss due to rice pests.[18] To remove the rice pests, the farmers would need to synchronize the flooding of their paddy fields to prevent the pests from breeding. However, this is not always possible due to the limitation of the water resource. The competitive requirements of balancing between the necessity to reduce rice pest locally and the obligation of conserving water globally have led to a system that is frustrated. We surmise that as the farmers strive to optimize their rice harvest, they have self-organized themselves into a critical state. Such a state turns out to offer a level of Pareto-optimality to all the farmers in the Balinese subaks.

In order to understand how the farmers attain Pareto-optimality through the process of self-organized criticality, we developed a simple toy model.[19]

In this model, a farmer would choose his planting phase based on one of four cropping schedules: flooded, drained, growing rice, and harvest. As before, if the farmers and its neighbors were to synchronize their planting phase, rice pests would drop while water stress increased. On the other hand, if they do not synchronize their planting phase, water stress will diminish at the cost of an increase in pest damage. The costs of these factors were embedded in the harvest function which each farmer computed at each point of time. After each computation, the farmer checked the harvest of its neighbors and then adopted the schedule of the neighbor that achieved the highest harvest. By performing this evaluation for all farmers, we observed an evolution of the cropping patterns of the subak towards a level of harvest that approached Pareto-optimality.

The observed cropping pattern depends on the relative weight between pest stress and water stress accorded by the subaks. If the subak farmers gave more weight to water stress, the cropping pattern would have many small patches which would save on water but encourage the spreading and breeding of pests from the neighboring fields. Conversely, if more weights were assigned to pest stress, we observed relatively large patches in the cropping pattern as fallowing large area would kill pests at the price of using more water. Interestingly, empirical data from satellite imagery displays neither of these cropping patterns. In other words, empirical data indicates that the cluster size of the cropping pattern does not favor either a predominantly large number of small patches or a majority of large cropping patches as described above. Instead, the data shows that the cluster size of the cropping patterns of each subak follows a power-law distribution. The subaks thus selected a state which balances the weight of the pest stress and water stress according to our toy model, such that the cropping pattern has the form of a power-law for the cluster size. This selection was made in a self-organized way by each subak as it adapted to its unique social-environmental conditions. Hence, we term the underlying process that drives the evolution of the observed cropping patterns *adaptive self-organized criticality*. We further affirmed that the subak system is indeed in a critical state by examining the correlation length of the cropping patches. Both model and data show that the correlation length exhibits a power-law behavior which indicates the presence of long-range correlation, or system-wide-connectivity of patches.

Next, we generalize the social behavior and decision mechanism that result in the observed cropping patterns through a Hamiltonian approach.[20] We propose a subak Hamiltonian of the following form:

$$H_S = -a \sum_{\langle i,j \rangle} \delta\left(\sigma_i, \sigma_j\right) + \frac{b}{N-1} \sum_{i>j} \delta\left(\sigma_i, \sigma_j\right), \tag{4}$$

where σ_i represents one of the four cropping schedules at site i of a 2D square lattice of linear size L, $\langle i, j \rangle$ means the sum is to be taken over the four neighboring sites, and $\delta(\cdot)$ is the Kronecker delta. The first sum corresponds to the effect of pest stress weighted by the parameter a. Because pest stress is determined by whether neighboring subaks adopt the same cropping schedule, it acts like a ferromagnetic Potts model with local interaction between nearest neighbors. The first sum thus promotes ordering at low temperature, where temperature effect captures the outcome of farmers in subaks not following the social mechanisms or irrigation schedules agreed upon in subaks meetings. On the other hand, the second sum involves all pairs of different sites on the lattice, with N being the total number of lattice sites, i.e. $N = L \times L$. This sum models the limitation of the water supply whose effect is regulated by the parameter b. It is a global or system-wide interaction without the usual distance attenuation factor which frequently appears in long-range interactions. It has a factor of $1/(N-1)$ to normalize the overall global interactions to ensure that H_s is an extensive quantity. With a positive sign, this sum implies a long-range antiferromagnetic contribution. It has the form of an antiferromagnetic Curie–Weiss Potts model and it serves to balance the local contribution of the ferromagnetic Potts model.

As before, interesting physics happens at the regime where pest stress balances water stress. This occurs when the weights of pest stress and water stress are equal in value. Our simulation of the subak Hamiltonian under this condition using Metropolis importance sampling Monte Carlo show the presence of two structural phase transitions. At low temperature, there is a competition at the boundary of the patches in maintaining a short (straight) boundary due to local interaction, versus the development of curvature at the boundary between the patches as a consequence of global interaction. If local interaction dominates, we observe the emergence of a phase of four equal-sized patches, which we call the *quadrant phase* based on simulation of the simple toy model covered earlier in this section. As the effect of global interaction becomes more apparent when temperature increases, four slightly unequal sized patches appear. These are meta-stable states and there are a myriad of them, reminiscent of the phenomenon of frustration in spin glasses. The first phase transition is a transition from the quadrant phase to a phase of four cropping patches of slightly different size. The latter phase is robust and exists for an extended temperature range before it is disrupted by entropic effect in an order–disorder phase transition. At the critical temperature of this second phase transition, we observe clusters of cropping patches of different sizes obeying a power-law distribution like the case of an adaptive SOC described earlier. We deduce the phase transition here to be second-order, and if not, at least weak first-order.

Finally, let us look at the Hamiltonian, i.e. Eq. (4) again but with σ_i taking two values (+1 and −1, like a spin) instead of four values:

$$H_S = -a \sum_{\langle i,j \rangle} \sigma_i \sigma_j + \frac{b}{N-1} \left(\sum_{i=1}^{N} \sigma_i \right)^2. \tag{5}$$

In this formulation, the second sum gives a clearer illustration of the global constraint of water conservation, exemplified by the benefits of an egalitarian distribution of water to the two different cropping schedules. Expanding this second sum gives rise to

$$\frac{b}{N-1} \left(\sum_{i=1}^{N} \sigma_i \right)^2 = \frac{b}{N-1} \left(\sum_{i=1}^{N} \sigma_i^2 + \sum_{i \neq j}^{N} \sigma_i \sigma_j \right) = \frac{b}{N-1} \sum_{i \neq j}^{N} \sigma_i \sigma_j + \frac{bN}{N-1},$$

where we have used the fact that $\sum_{i=1}^{N} \sigma_i^2 = \sum_{i=1}^{N} 1 = N$. Because a constant term is irrelevant in a Hamiltonian, Eq. (5) becomes

$$H_S = -a \sum_{\langle i,j \rangle} \sigma_i \sigma_j + \frac{b}{N-1} \sum_{i \neq j}^{N} \sigma_i \sigma_j. \tag{6}$$

The second term in the above equation describes the Curie–Weiss all-to-all antiferromagnet which appears in Eq. (4).

Using the formulation of Eq. (6), let us consider three scenarios. For the first scenario where $a \gg b$, we have

$$H_s \approx -a \sum_{\langle i,j \rangle} \sigma_i \sigma_j ,$$

which indicates that the system behaves like an Ising ferromagnet. In this case, we observe the expected large cropping patches since pest stress dominates. At the other extreme where $b \gg a$, we have the Curie–Weiss antiferromagnet:

$$H_s \approx \frac{b}{N-1} \sum_{i \neq j}^{N} \sigma_i \sigma_j ,$$

with the cropping pattern displaying many small patches. When $a \approx b$, Eq. (6) could be likened to a spin-glass Hamiltonian (even though a and b are not random) with the given values of a and b treated as one random realization. Below the critical temperature, we expect a multitude of "frozen" states while above the critical temperature, one single paramagnetic state. Nonetheless, there exist a ground state of two equal size large patches of different cropping schedules — the equivalent of the quadrant phase of the four-state subak Hamiltonian. As the system equilibrates towards this ground state, it typically gets stuck in some local minima. The subak Hamiltonian thus describes a frustrated and nondegenerate system, with a rough funnel energy landscape similar to the protein folding problem. This perspective is shared by a cellular Potts model, whose Hamiltonian of interfacial energy, area conservation, and motility contains rather similar physics to our subak Hamiltonian.[21] For the cellular Potts model, there is a dynamical transition from a fluid phase to a solid phase, where glassy behaviors occur in the solid phase as subdiffusion, very slow relaxation, and aging. For our case, glassy behavior happens below the critical temperature as the system relaxes very slowly from the multitude of meta-stable states towards the ground state.

5. Conclusion

Unlike phase transition in typical systems, the phase transition in complex systems is glassy and contains a multitude of meta-stable states and these states could be frustrated. This conforms to the idea that a complex system is diverse, history dependent, and in a state of nonequilibrium. We observe these characteristics in protein phase transition, and in a coupled social-ecological system that obeys a Hamiltonian generalized from the social behavioral and decision mechanisms of Balinese subaks. The social-ecological system also displays self-organized criticality which is metaphorically modeled by the sandpile model. In our context, the farmers optimizes their harvests by adapting to the environmental conditions that balance pest and water stress. In so doing, the system converges to a state of criticality in a self-organized and adaptive way, which we modeled by choosing the a and b values that balance the stresses in the Hamiltonian model. One important consequence of a system in criticality is the presence of universality class whose membership is determined by the values of scaling exponents that obey a certain set of scaling laws. We studied one such model in the Manna universality class (the Abelian Manna model) by examining the model's scaling behavior as it undergoes self-organized criticality in fractal lattices. One would question whether such scaling laws and scaling exponents could occur in the coupled social-ecological system that we studied. While scaling law and exponent have been found for protein system relating the radius of gyration and number of amino acid residues, as well as in self-organized critical systems from Hamiltonians with glassy dynamics,[22–24] the presence of scaling laws, scaling exponents, and universality class in social systems is as yet unexplored. This outstanding problem would be an interesting area to pursue in our future work.

References

1. D. L. Stein and C. M. Newman, *Spin Glasses and Complexity* (Princeton University Press, New Jersey, 2013).
2. J. D. Bryngelson and P. G. Wolynes, Spin glasses and the statistical mechanics of protein folding, *Proc. Natl. Acad. Sci. USA* **84** (1987) 7524–7528.

3. S. S. Rao and S. M. Bhattacharjee, Protein folding and spin glass, *Physica A* **224** (1996) 279–286.

4. K. Huang, Conditioned self-avoiding walk (CSAW): Stochastic approach to protein folding, *Biophys. Rev. Lett.* **2**(2) (2007) 139–154.

5. J. Lei and K. Huang, Elastic energy of proteins and the stages of protein folding, *EPL* **88** (2009) 68004.

6. H. W. Leong, L. Y. Chew and K. Huang, Normal modes and phase transition of the protein chain based on the Hamiltonian formalism, *Phys. Rev. E* **82** (2010) 011915.

7. H. W. Leong and L. Y. Chew, Effect of hydrogen bond interaction on protein phase transition, *Phys. Rev. E* **86** (2012) 031902.

8. E. Callaway, What's next for the AI protein folding revolution, *Nature* **604** (2022) 234–238.

9. P. Bak, *How Nature Works: The Science of Self-organized Criticality* (Springer-Verlag, New York, 1996).

10. K. Christensen and N. R. Moloney, *Complexity and Criticality* (Imperial College Press, London, 2005).

11. R. Dickman, A. Vespignani and S. Zapperi, Self-organized criticality as an absorbing-state phase transition, *Phys. Rev. E* **57** (1998) 5095.

12. P. Bak, C. Tang and K. Wiesenfeld, Self-organized criticality: An explanation of the $1/f$ noise, *Phys. Rev. Lett.* **59** (1987) 381.

13. A. Petri and G. Pontuale, Morphology and dynamics in SOC universality classes, *J. Stat. Mech.* (2018) 063201.

14. H. N. Huynh, L. Y. Chew and G. Pruessner, Abelian Manna model on two fractal lattices, *Phys. Rev. E* **82** (2010) 042103.

15. H. N. Huynh, G. Pruessner and L. Y. Chew, The Abelian Manna model on various lattices in one and two dimensions, *J. Stat. Mech.* (2011) P09024.

16. D. Plenz, T. L. Ribeiro, S. R. Miller, P. A. Kells, A. Vakili and E. L. Capek, Self-organised criticality in the brain, *Front. Phys.* **9** (2021) 639389.

17. J. S. Lansing, *Perfect Order: Recognizing Complexity in Bali* (Princeton University Press, New Jersey, 2006).

18. H. S. Sugiarto, J. S. Lansing, N. N. Chung, C. H. Lai, S. A. Cheong and L. Y. Chew, Social cooperation and disharmony in communities mediated through common pool resource exploitation, *Phys. Rev. Lett.* **118** (2017) 208301.

19. J. S. Lansing, S. Thurner, N. N. Chung, A. Coudurier-Curveur, C. Karakas, K. A. Fesenmyer and L. Y. Chew, Adaptive self-organisation of Bali's ancient rice terraces, *Proc. Natl. Acad. Sci. USA* **114**(25) (2017) 6504–6509.

20. Y. Gandica, J. S. Lansing, N. N. Chung, S. Thurner and L. Y. Chew, Bali's ancient rice terraces: A Hamiltonian approach, *Phys. Rev. Lett.* **127** (2021) 168301.

21. M. Chiang and D. Marenduzzo, Glass transitions in the cellular Potts model, *EPL* **116** (2016) 28009.
22. J. C. Andresen, Z. Zhu, R. S. Andrist, H. G. Katzgraber, V. Dobrosavljević and G. T. Zimanyi, Self-organised criticality in glassy spin systems requires a diverging number of neighbors, *Phys. Rev. Lett.* **111** (2013) 097203.
23. F. Pázmándi, G. Zaránd and G. T. Zimanyi, Self-organised criticality in the hysteresis of the Sherrington-Kirkpatrick model, *Phys. Rev. Lett.* **83** (1999) 1034.
24. J. C. Phillips, Scaling and self-organised criticality in proteins I, *Proc. Natl. Acad. Sci. USA* **106**(9) (2009) 3107–3112.

Masses of Weak Bosons and Dark Matter Particles

Harald Fritzsch

Department für Physik,
Ludwig-Maximilians-Universität, München, Germany

I first describe my relations with Kok Khoo Phua and my stays at the IAS in Singapore.

The masses of the three weak bosons are generated in a similar way as the proton mass in QCD. The three weak bosons are bound states of two fermions and their antiparticles. There exist also two bound states of three fermions. One of them is neutral and stable. This particle provides the dark matter in our universe.

In 1978 I was a professor at the university of Bern in Switzerland. In January 1978, I received a letter from the organizing committee of the XIX International Rochester Conference, which was scheduled to be held in August 1978 in the Keio Plaza Hotel in Tokyo. I was invited to give a lecture on the electroweak interactions.

Then a letter arrived from Prof. Phua, who was at this time a professor at the National University of Singapore. I was invited to give a lecture at a conference, which was planned to be held just before the Rochester Conference at the RELC hotel in Singapore.

I decided to accept the invitations to Tokyo and Singapore, since Singapore is not very far away from Japan. At the conference of Prof. Phua, I gave the same lecture as at the Rochester Conference later.

In Singapore, I met Prof. Kok Khoo Phua for the first time. Singapore in 1978 was much different from the Singapore today — the current average lifetime of a building in Singapore is only about 15 years.

The RELC hotel is close to Orchard Road, which is the central street in Singapore and where there are many restaurants. On Orchard Road, I enjoyed lunches with Prof. Phua, or with John Sakurai or Leo Stodolsky.

Thirty years later, in November 2008, I was again in Singapore at a conference, organized by Prof. Phua. He asked me to organize a big conference in 2010, on the occasion of the 80th birthday of Murray Gell-Mann. I should mention, that I knew Gell-Mann very well. From 1970 until 1975, I had collaborated with him. In 1972 we had found the correct theory of the strong interactions, which we called Quantum Chromodynamics and which is today part of the Standard Theory of particle physics. The conference for Gell-Mann took place in February 2010 at the Nanyang Technological University in Singapore.

Prof. Phua created a new institute at NTU, the Institute of Advanced Studies (IAS). I visited this institute every year, from 2012 until 2018, usually for a period of two months. I organized four big conferences, in 2015, 2016, 2017 and 2018, for example in 2016 the International Conference on New Physics at the Large Hadron Collider. In 2018, Prof. Phua and his wife Doreen visited us in Munich, and we had dinner in a good restaurant.

Now I come to the physics. In my office at the IAS in Singapore I was considering a possible internal structure of the three weak bosons. In the Standard Theory of particle physics the massive weak bosons are point-like and obtain their masses by the spontaneous breaking of the gauge symmetry. I do not like this kind of mass generation. The concept of mass is a fundamental concept and there should be a dynamical mechanism to generate all masses in physics, the proton mass, the masses of the leptons and quarks and the masses of the weak bosons.

The masses of the bound states in confining gauge theories are generated by a mechanism, which is due to the confinement property of the non-Abelian gauge bosons. For example, in the theory of Quantum Chromodynamics the masses of the hadrons can be calculated. They are given by the field energy of the confined gluons and quarks and are proportional to the QCD mass scale Λ_c, which has to be measured in the experiments: $\Lambda_c = 332 \pm 17$ MeV.

In the Standard Theory, the physics of the six leptons and the six quarks depends on 20 free parameters: twelve masses of the leptons and quarks and eight parameters of the flavor mixing — six mixing angles and two phase parameters, which describe the CP violation.

These twenty parameters might indicate, that the leptons, quarks and weak bosons are not point-like, but composite systems. Especially we shall assume, that inside the leptons and quarks are two point-like particles, a fermion and a scalar. There are two fermions and four scalars — they are called "haplons" ("haplos" means "simple" in the Greek language).

The haplons are confined by massless gauge bosons, which are described by a gauge theory similar to Quantum Chromodynamics. We call these gauge bosons "hypergluons" and the associated quantum numbers "hyper-colors". The new gauge theory is called "Quantum Haplodynamics" (QHD). I shall not discuss the internal structure of the leptons and quarks in more detail, but move now to the weak bosons.

We assume that the masses of the weak bosons are also generated dynamically. This is possible, if the weak bosons are not elementary gauge bosons, but bound states of haplons, analogous to the ρ-mesons in QCD.

The weak bosons consist of two haplons, a fermion and its antiparticle (see also Refs. 1–5). The QHD mass scale is described by a mass parameter Λ_h, which determines in particular the size of the weak bosons.

The haplons are massless and interact with each other through the exchange of massless hypergluons. The number of hypergluons depends on the gauge group, which is unknown. We assume, that it is the same as the gauge group of QCD: SU(3). Thus the binding of the haplons is due to the exchange of eight massless hypergluons.

Two types of fermions are needed as constituents of the weak bosons, denoted by α and β. The electric charges of these two haplons are (+2/3) and (−1/3). Before the mixing of the neutral weak boson and the photon, the three weak bosons have the same mass and the following internal structure:

$$W^+ = (\bar{\beta}\alpha), \quad W^- = (\bar{\alpha}\beta), \quad W^3 = \frac{1}{\sqrt{2}}(\bar{\alpha}\alpha - \bar{\beta}\beta). \tag{1}$$

The QHD mass scale can be estimated, using the observed value of the decay constant of the weak boson. This constant is analogous to the decay constant of the ρ-meson, which is directly related to the QCD mass scale. The decay constant of the weak boson is given by the mass of the weak boson and the observed value of the weak mixing angle:

$$F_W = \sin\theta_W \cdot \frac{M_W}{e} . \tag{2}$$

We find $F_W \simeq 0.124$ TeV. Thus we expect, that the QHD mass scale Λ_h is in the range between 0.2 TeV and 1 TeV. The ratio of the QHD mass scale and of the QCD mass scale is about three orders of magnitude.

There should exist also fermions, which consist of three haplons. There are two such particles:

$$(\alpha\alpha\beta) = Y^+, \ (\alpha\beta\beta) = Y^0 . \tag{3}$$

If the electromagnetic interaction is neglected, these two fermions would have the same mass. After introducing electromagnetism, this degeneracy is lifted. The electromagnetic field around the charged Y-particle produces a mass. We have in particular:

$$m(Y^+) = m(Y^0) + \delta m . \tag{4}$$

Let us assume, that the mass of the Y^0-particle is 300 GeV. In this case the electromagnetic massshift δm is about 200 MeV:

$$m(Y^+) \simeq 300.2 \text{ GeV} . \tag{5}$$

The charged Y-particle decays into the neutral one via the weak interaction, for example $Y^+ \to Y^0 + \pi^+$.

The lifetime of the charged Y-particle is about 10^{-9} seconds. The neutral Y-particle consists of three haplons. It is stable, since the haplon number is conserved.

In proton–proton or antiproton–proton collisions, the Y-particles must be produced in pairs, e.g. $p + p \to Y^+\overline{Y^+} + \ldots$ We consider now the production of a Y^+-particle and its antiparticle. This particle decays into a Y^0-particle by emitting a virtual positively charged weak boson:

$$Y^+ \to Y^0 + {}``W^+" \tag{6}$$

For example, the charged Y-particle can decay into the neutral one by emitting a charged pion. The Y^+-particle decays inside the detector of the LHC. The track of this particle can probably be observed. The length of the track might be 0.5 m. In such an event, one could also observe the decay products of the two virtual weak bosons, for example, two pions. The Y^0-particle or its antiparticle cannot be directly observed, but indirectly due to the missing energy.

During the Big Bang the Y-fermions and their antiparticles would have been produced, as well as quarks and antiquarks or leptons and their antiparticles. Since there is a small asymmetry between the matter and the antimatter in the universe, due to the violation of the CP-symmetry, there are more quarks than antiquarks. The antiquarks disappear — they annihilate with the quarks and produce photons. The remaining quarks are the constituents of the matter in our universe.

Due to CP-violation there are also more Y^0-particles than $\overline{Y^0}$-particles, or more $\overline{Y^0}$-particles than Y^0-particles. Those would annihilate with the Y^0-particles or $\overline{Y^0}$-particles, and finally the universe would contain, besides protons and neutrons, a gas of stable Y^0-particles or $\overline{Y^0}$-particles, providing the dark matter in our universe. It is not possible to calculate, how much dark matter exists — this would imply that one can calculate the density of the Y^0-particles or $\overline{Y^0}$-particles, which is not possible.

The average density of the dark matter in our galaxy has been measured to be about 0.39 GeV/cm^3. If the mass of the neutral Y-particle is 300 GeV, there would be about 1 300 Y^0-particles or $\overline{Y^0}$-particles in one cubic meter. Our galaxy is inside a big cloud of dark matter. In this cloud the neutral Y-particles move with an average speed of about 15 m/sec.

The Sun moves around the center of our galaxy with a speed of about 200 km/s. The speed of the Earth around the Sun is about 31 km/s. Thus the speed of the Earth inside the dark matter cloud varies between 169 km/s and 231 km/s — the neutral Y-particles here on Earth have a rather large speed.

A neutral Y-fermion can emit a virtual Z-boson, which interacts with an atomic nucleus. The cross-section for the reaction of a neutral Y-fermion with a nucleus should be about 10^{-45} cm^2. In a collision of a neutral Y-fermion with a nucleus, one can observe in specific experiments, the sudden change of the momentum of the nucleus. Thus the neutral Y-fermions can be observed indirectly, e.g. in the experiments in the Gran Sasso Laboratory, but the relevant cross-section might be too small.

With the Large Hadron Collider (LHC) the Y-particles are produced by the direct interaction of a quark and an antiquark. They cannot be produced by the interaction of two gluons. Thus the production cross-section would be larger, if the LHC is a proton–antiproton collider. But even with the present LHC, I expect that the charged Y-fermions can be observed soon.

References

[1] H. Fritzsch and G. Mandelbaum, *Phys. Lett. B* **102** (1981) 319; *Phys. Lett. B* **109** (1982) 224.
[2] R. Barbieri, R. Mohapatra and A. Masiero, *Phys. Lett. B* **105** (1981) 369.
[3] H. Fritzsch, D. Schildknecht and R. Kogerler, *Phys. Lett. B* **114** (1982) 157.
[4] L. F. Abbott and E. Farhi, *Phys. Lett. B* **101** (1981) 69.
[5] H. Fritzsch, *Mod. Phys. Lett. A* **31**(20) (2016) 1630019.

FXGAR Mass Matrices: A Unique Viable Set for Quarks and Leptons

生
日
快
乐

Manmohan Gupta[†], Gulsheen Ahuja[‡], Aakriti Bagai,
Mahak Garg, Aseem Vashisht and Kanwaljeet S. Channey[*]

Department of Physics, Panjab University, Chandigarh, India
[*]Department of Physics, University Institute of Sciences, Chandigarh University,
Punjab, India
[†]mmgupta@pu.ac.in
[‡]gulsheen@pu.ac.in

FXGAR mass matrices (to be defined later) are a particular class of texture 4 zero hermitian mass matrices, known to be quite successful in explaining the quark mixing data. Following the spirit of quark lepton symmetry as well as assuming neutrinos to be Dirac particles, an attempt has been made to examine the viability of FXGAR mass matrices for the lepton sector. Our analysis reveals that Inverted Ordering (IO) of neutrino masses is ruled out while Normal Ordering (NO) continues to be viable.

To Prof. K.K. Phua

This present manuscript being a part of the volume published as a tribute to Prof. K.K. Phua on his 80th birthday, therefore, it would in the fitness of things to add few words from our side to emphasize the role played by him in shaping our ideas and approach towards research as well as of many young researchers in all parts of the world. Our exposure to conferences organized at IAS@NTU started with the one, 'Particle Physics, Astrophysics and Quantum Field Theory: 75 Years since Solvay', organized by Prof. Phua in November 2008. After this conference, almost every year, conferences or workshops were organized under the patronage of Prof. Phua who was very generous in extending the invitation to young researchers as well in hospitality. This attracted large numbers of young researchers all over the world. The conferences always had participation by couple of Nobel laureates as well as by extremely distinguished scientists from across the globe. These conferences/workshops, on the one hand, gave opportunity to young researchers from all over the world, particularly from Asia, to have the opportunity to learn from the best in the field, on the other hand , provided a unique opportunity to push their out of the box ideas through the quality journals of Singapore Scientific Publications.

It would be desirable to mention that our visits to IAS@NTU, Singapore on several occasions proved instrumental in shaping our ideas in this field. Two of us (MG and GA) attended almost all the conferences/workshops, organized since 2008, by Prof. Phua at IAS@NTU Singapore, with MG having the privilege of spending more than a month in 2011 at IAS on the invitation of Prof. Phua. These visits not only provided us the opportunity to share our ideas with Harald Fritzch, N. P. Chang, among many other distinguished experts, but also, facilitated the publication of these in the prestigious WSPC journals, including two review articles. The invitation of the publication for a long review article, 'Flavor mixings and texture of the fermion mass matrices' [30], motivated us to explore the field with greater enthusiasm. This resulted into several high quality research papers in the field. To say the least, Prof. Phua's generosity played a great role in providing exposure and visibility to us. The work presented here is in continuation of our efforts to understand the so called 'Flavor Riddle' with its motivation lying in the good number of conferences we attended at IAS@NTU.

1. Introduction

Unravelling the 'Flavor Puzzle', in other words, understanding fermion masses and mixings is one of the main objectives of present day research in High Energy Physics for both experimentalists and theoriticians. On the experimental front, in the last few years, remarkable progress has been made in the measurements of the fermion masses and mixing parameters. In particular, for the quark sector, most of the Cabibbo Kobayashi Maskawa (CKM) parameters [1,2] are now known within an accuracy of 10% or less [3]. Also, there has been considerable sharpening of data regarding the quark masses, in particular of those of the light quark masses, m_u, m_d and m_s [4]. Similarly, in the context of neutrino oscillation phenomenology, impressive progress has been made in the measurement of the neutrino masses and mixing parameters through solar [5–11], atmospheric [12–14], reactor (KamLAND [15], Double-Chooz [16], Daya-Bay [17], Reno [18]), and accelerator (MINOS [19], T2K [20], NOvA [21]) neutrino experiments. Therefore, one may say that refinements of the parameters in both the quark and lepton sectors can now be considered to be at the stage of 'precision measurements' and would have far reaching consequences for Flavor Physics. In view of the relationship of the fermion mass matrices with the corresponding fermion mixing matrices, developments pertaining to refinements of the fermion mixing data would undoubtedly have deep implications for the structure of the mass matrices.

In order to achieve a theoretical understanding of fermion masses and mixings, one may follow either of the two approaches, i.e. 'top-down' [22–25] and 'bottom-up' [26–30]. In the absence of any viable, universally accepted, model from the top-down approach, the emphasis is on the bottom-up approach for finding a viable set of fermion mass matrices. Proceeding along the bottom-up approach, one tries to find the phenomenological fermion mass matrices which are in tune with the low energy data and can serve as guiding stone for developing more ambitious theories. In the Standard Model (SM), the fermion mass matrices appear as completely arbitrary complex matrices with a total of 36 real free parameters, whereas, the number of physical observables are much smaller, e.g., 10 in the quark sector, with a similar number in the leptonic sector for Dirac neutrinos (2 additional ones in case neutrinos are Majorana particles). Therefore, as a first step, in order to develop viable phenomenological fermion mass matrices, one has to limit the number of free parameters in the mass matrices. To this end, it should be noted that in the SM and its extensions, a rotation of the right-handed fermion fields does not affect any physical results, hence, the mass matrices can be considered as hermitian without loss of generality, immediately bringing down the number of real free parameters. This number can be reduced further, by using the idea of textures zero mass matrices, initiated implicitly by Weinberg [31] and explicitly by Fritzsch [32,33], wherein some of the entries of the mass matrices are assumed to be zero.

A particular texture 'n' zero mass matrix is defined such that sum of the number of zeros at diagonal positions and a pair of symmetrically placed zeros at off-diagonal positions, counted as 1, is 'n'. The original Fritzsch ansatz of hermitian quark mass matrices is given by [32,33]

$$M_U = \begin{pmatrix} 0 & A_U & 0 \\ A_U^* & 0 & B_U \\ 0 & B_U^* & C_U \end{pmatrix}, \quad M_D = \begin{pmatrix} 0 & A_D & 0 \\ A_D^* & 0 & B_D \\ 0 & B_D^* & C_D \end{pmatrix}, \tag{1}$$

where M_U and M_D correspond to mass matrices in the up (U) and down (D) sector. Both the matrices M_U and M_D are texture 3 zero type, together this set of matrices is referred to as texture 6 zero mass matrices. An immediate extension of the above mentioned texture is the texture 5 zero mass matrices which can be obtained by replacing one of the zero entry in either of the texture 3 zero matrices with a non-zero one. For example, considering the (2,2) element of any of the above mentioned matrices M_U and M_D to be non-zero, one obtains the following texture 2 zero mass matrix

$$M_i = \begin{pmatrix} 0 & A_i & 0 \\ A_i^* & D_i & B_i \\ 0 & B_i^* & C_i \end{pmatrix}, \tag{2}$$

where $i = U, D$ and D_i is the real element of the matrix. Along with these matrices, depending upon the position of zeros there are several other possible structures which can be considered to be texture 2 zero ones. Thereafter, one arrives at texture 5 zero mass matrices by considering either of the mass matrix in the up or the down sector to be texture 3 zero type, e.g. the one given in Eq. (1), along with the mass matrix in the other sector being 2 zero type, e.g. the one given in Eq. (2). As a next step, the texture 4 zero combinations can be obtained when one considers both M_U and M_D to be texture 2 zero type.

In the quark sector, keeping in mind the recent precision measurements of light quark masses, the idea of texture zero mass matrices for all possibilities of texture 6, 5 and 4 zero mass matrices has been recently explored in detail [34–36]. In particular, it has been shown [34] that the present refined data of quark masses and mixing angles unambiguously rules out the original Fritzsch ansatze, mentioned in Eq. (1), as well as all of its texture 6 zero variants. Further, the present data also rules out all possible texture 5 zero hermitian mass matrices [35]. In the case of texture 4 zero mass matrices, out of all possibilities compatible with Weak Basis transformations [37], a particular set of mass matrices, given in Eq. (2), along with its permutations, emerges to be a unique possibility which has the best possible compatibility with the quark mixing data. This set of mass matrices, being closely related to the original Fritzch ansatze, has been examined largely by Xing, Gupta, Ahuja and Randhawa [26–30,38], therefore, in the present work, this set of mass matrices would be referred to as FXGAR(Fritzch-Xing-Gupta-Ahuja-Randhawa) mass matrices.

In view of the possibility of the existence of an almost unique, viable set of quark mass matrices, following the spirit of quark lepton symmetry [39], it becomes interesting to carry out a similar exercise for this set in the case of leptons as well. Before proceeding further, it should be borne in mind that at present it is not clear whether neutrinos are Dirac or Majorana particles. While a positive signal from neutrinoless double beta decay ($0\nu\beta\beta$) experiments can confirm the Majorana nature of light neutrinos, yet, at present, we do not have any such signal with latest data from KamLAND-Zen experiment [40]. Although an absence of a signal from neutrinoless double beta decay experiments does not necessarily rule out the Majorana nature of light neutrinos, yet, it provides a good deal of motivation to keep the option of neutrinos being light Dirac particles open.

In the last several years several Beyond the Standard Model (BSM) scenarios have been proposed in order to explain light Dirac neutrino masses [41–47]. In most of these models, Majorana mass of right-handed neutrinos are forbidden due to additional symmetries and some of these also explain the natural origin of tiny Dirac neutrino masses via some kind of seesaw mechanism. Motivated by the growing interests in Dirac neutrinos, several analyses have been carried out in the case of texture specific mass matrices

[48–53]. In the light of several precision measurements pertaining to neutrino mixing parameters over the last few years, in the present work, we have investigated FXGAR mass matrices for the case of leptons. in particular, we have examined the viability of these for the two possible scenarios of neutrino mass orderings, i.e. NO and IO. Further, corresponding to the viable possibility, we have made an attempt to determine ranges of the lightest neutrino mass and the Jarlskog's rephasing invariant parameter in the leptonic sector.

The paper is organized as follows. To begin with, in Sec. 2, after defining FXGAR mass matrices for Dirac neutrinos, we present essentials of the methodology connecting the mass matrices to the lepton mixing matrix. Inputs used in the analysis are given in Sec. 3. Results and discussion pertaining to the compatibility of the mass matrices are presented in Sec. 4. Finally, Sec. 5 summarizes our conclusions.

2. FXGAR Mass Matrices for the Case of Leptons

FXGAR mass matrices for the case of leptons are given by

$$M_l = \begin{pmatrix} 0 & A_l & 0 \\ A_l^* & D_l & B_l \\ 0 & B_l^* & C_l \end{pmatrix}, \quad M_{\nu_D} = \begin{pmatrix} 0 & A_{\nu_D} & 0 \\ A_{\nu_D}^* & D_{\nu_D} & B_{\nu_D} \\ 0 & B_{\nu_D}^* & C_{\nu_D} \end{pmatrix}, \tag{3}$$

where M_l and M_{ν_D} correspond to the mass matrices for the charged leptons and Dirac neutrinos respectively, with complex off diagonal elements, i.e. $A_i = |A_i|e^{i\alpha_i}$ and $B_i = |B_i|e^{i\beta_i}$, where $i = l, \nu_D$, whereas C_i and D_i are the real elements of the matrix. In parallel with the methodology to construct mixing matrix from mass matrices in the quark sector, to begin with, we consider a general mass matrix in the leptonic sector M_i ($i = l, \nu_D$) as

$$M_i = P_i^\dagger M_i^r P_i, \tag{4}$$

where M_i^r is real matrix and P_i denotes the phase matrix given as $\text{Diag}(e^{i\alpha_i}, 1, e^{-i\beta_i})$. The real matrix M_i^r can then be diagonalized by the orthogonal transformations O_i, i.e.

$$M_i^{\text{Diag}} = O_i^T M_i^r O_i \tag{5}$$

which can be rewritten as

$$M_i^{\text{Diag}} = O_i^T P_i M_i P_i^\dagger O_i = U_i^\dagger M_i U_i,$$ (6)

where $O_i^T P_i = U_i^\dagger$ and $P_i^\dagger O_i = U_i$ denotes diagonalizing unitary matrices. $M_i^{\text{Diag}} = \text{Diag}(m_1, -m_2, m_3)$, where the subscripts 1, 2 and 3 refer respectively to e, μ and τ for the charged leptons and ν_1, ν_2 and ν_3 for the neutrinos. The diagonalizing transformation O_i for the matrices given in Eq. (3) is given as

$$O_i = \begin{pmatrix} \sqrt{\dfrac{m_2 m_3 (C_i - m_1)}{C_i(m_1+m_2)(m_3-m_1)}} & \sqrt{\dfrac{m_1 m_3 (C_i + m_2)}{C_i(m_1+m_2)(m_3+m_2)}} & \sqrt{\dfrac{m_1 m_2 (m_3 - C_i)}{C_i(m_3+m_2)(m_3-m_1)}} \\[2.5ex] \sqrt{\dfrac{m_1 (C_i - m_1)}{(m_1+m_2)(m_3-m_1)}} & -\sqrt{\dfrac{m_2 (C_i + m_2)}{(m_1+m_2)(m_3+m_2)}} & \sqrt{\dfrac{m_3 (m_3 - C_i)}{(m_3-m_1)(m_2+m_3)}} \\[2.5ex] -\sqrt{\dfrac{m_1 (C_i + m_2)(m_3 - C_i)}{C_i(m_1+m_2)(m_3-m_1)}} & \sqrt{\dfrac{m_2 (m_3 - C_i)(C_i - m_1)}{C_i(m_1+m_2)(m_3+m_2)}} & \sqrt{\dfrac{m_3 (C_i - m_1)(C_i + m_2)}{C_i(m_3+m_2)(m_3-m_1)}} \end{pmatrix}.$$ (7)

The corresponding diagonalizing matrix for the charged leptons, i.e. the matrix O_l can be obtained by replacing m_1, $-m_2$, m_3 with m_e, $-m_\mu$ and m_τ, in the above matrix, e.g. the first element of O_l can be given as

$$O_{11} = \sqrt{\frac{m_\mu m_\tau (C_l - m_e)}{C_l (m_e + m_\mu)(m_\tau - m_e)}}.$$ (8)

One can obtain all other elements of the O_l in a similar way. To obtain the diagonalizing matrix for Dirac neutrinos, O_{ν_D}, for NO scenario, given by $m_{\nu_1} < m_{\nu_2} \ll m_{\nu_3}$, one can replace m_1, $-m_2$, m_3 in Eq. (7) with m_{ν_1}, $-m_{\nu_2}$, m_{ν_3} respectively. For instance, first element of O_{ν_D} can be given as

$$O_{11} = \sqrt{\frac{m_{\nu_2} m_{\nu_3} (C_{\nu_D} - m_{\nu_1})}{C_{\nu_D} (m_{\nu_1} + m_{\nu_2})(m_{\nu_3} - m_{\nu_1})}}.$$ (9)

For the IO case, defined as $m_{\nu_3} \ll m_{\nu_1} < m_{\nu_2}$, the elements of diagonalizing transformation can be obtained by replacing m_1, $-m_2$, m_3 with m_{ν_1}, $-m_{\nu_2}$, $-m_{\nu_3}$ in Eq. (7), e.g. the first element of O_{ν_D} in this scenario can be given as

$$O_{11} = \sqrt{\frac{-m_{\nu_2} m_{\nu_3} (C_{\nu_D} - m_{\nu_1})}{C_{\nu_D} (m_{\nu_1} + m_{\nu_2})(-m_{\nu_3} - m_{\nu_1})}}.$$ (10)

The other elements of the orthogonal transformations for the neutrino mass matrix can be obtained in a similar manner.

The lepton mixing matrix, i.e. the Pontecorvo–Maki–Nakagawa–Sakata (PMNS) matrix [54] can be obtained from these orthogonal transformations using the relation

$$U_{\text{PMNS}} = O_l^T P_l P_{\nu_D}^\dagger O_{\nu_D} = U_l^\dagger U_{\nu_D},\qquad(11)$$

where U_l and U_{ν_D} are the unitary transformations for the matrices M_l and M_{ν_D} respectively. The product $P_l P_{\nu_D}^\dagger$ results in $\text{Diag}(e^{-i\phi_1}, 1, e^{i\phi_2})$, where ϕ_1 and ϕ_2 are related to the phases of mass matrices as $\phi_1 = \alpha_{\nu_D} - \alpha_l$ and $\phi_2 = \beta_{\nu_D} - \beta_l$.

3. Inputs Used for the Analysis

To carry out the present analysis, we use inputs given by Garcia et al. [55]. Corresponding to both NO and IO, the 1σ and 3σ ranges of the three mixing angles, the mass square differences and the likely CP violating phase δ have been presented in Table 1. It may be added that for the purpose of

Table 1. Three flavor oscillation parameters from global data [55]. Note that $\Delta m_{3\ell}^2 \equiv \Delta m_{31}^2 > 0$ for NO, while $\Delta m_{3\ell}^2 \equiv \Delta m_{32}^2 < 0$ for IO.

Parameter	Best Fit ± 1σ	3σ range
$\sin^2\theta_{12}$	$0.304^{+0.013}_{-0.012}$	$0.269 - 0.343$
$\theta_{12}[degrees]$	$33.44^{+0.78}_{-0.75}$	$31.27 - 35.86$
$\sin^2\theta_{23}(NO)$	$0.570^{+0.018}_{-0.024}$	$0.407 - 0.618$
$\theta_{23}[degrees]$	$49.0^{+1.1}_{-1.4}$	$39.6 - 51.8$
$\sin^2\theta_{23}(IO)$	$5.575^{+0.017}_{-0.021}$	$0.411 - 0.621$
$\theta_{23}[degrees]$	$49.3^{+1.0}_{-1.2}$	$39.9 - 52.0$
$\sin^2\theta_{13}(NO)$	$0.02221^{+0.00068}_{-0.00062}$	$0.02034 - 0.02430$
$\theta_{13}[degrees]$	$8.57^{+0.13}_{-0.12}$	$8.20 - 8.97$
$\sin^2\theta_{13}(IO)$	$0.02240^{+0.00062}_{-0.00062}$	$0.02053 - 0.02436$
$\theta_{13}[degrees]$	$8.61^{+0.12}_{-0.12}$	$8.24 - 8.98$
$\Delta m_{21}^2[\times10^{-5}\,eV^2]$	$7.42^{+0.21}_{-0.20}$	$6.82 - 8.04$
$|\Delta m_{3\ell}^2|[\times10^{-3}\,eV^2](NO)$	$+2.514^{+0.028}_{-0.027}$	$+2.431 - 2.598$
$|\Delta m_{3\ell}^2|[\times10^{-3}\,eV^2](IO)$	$-2.497^{+0.028}_{-0.028}$	$-2.583 - 2.412$
$\delta[degrees](NO)$	195^{+51}_{-25}	$107 - 403$
$\delta[degrees](IO)$	286^{+27}_{-32}	$192 - 360$

calculations pertaining to NO and IO of neutrino masses, the elements D_l, D_ν have been considered as free parameters, such that diagonalizing transformation O_l and O_{ν_D} remains real. Also, the phases ϕ_1 and ϕ_2 have been given full variation from 0 to 2π. The lightest neutrino mass, m_1 for NO and m_3 for IO, is taken to be 10^{-8} eV $- 10^{-1}$ eV, however, our conclusions remain unaffected even if the range is extended further.

The magnitude of the neutrino mixing matrix elements obtained by carrying out the global fit of experimentally measured parameters as given by Garcia *et al.* [55] is

$$U_{\text{PMNS}} = \begin{pmatrix} 0.801 - 0.845 \ 0.513 - 0.579 \ 0.143 - 0.156 \\ 0.233 - 0.507 \ 0.461 - 0.694 \ 0.631 - 0.778 \\ 0.261 - 0.526 \ 0.477 - 0.694 \ 0.613 - 0.756 \end{pmatrix} \qquad (12)$$

4. Analysis of FXGAR Mass Matrices for the Case of Leptons

For the case of Dirac neutrinos, as a first step, in order to examine the viability of FXGAR mass matrices, we evaluate the corresponding PMNS matrix and verify its compatibility with one obtained through global fits. Using the inputs mentioned in the previous section, we have carried out the analysis for the two neutrino mass orderings.

4.1. *Inverted ordering of neutrino masses*

To begin with, we carry out the analysis for the inverted ordering of neutrino masses. From the relation given in Eq. (11) and using the above mentioned inputs, we arrive at the following PMNS matrix

$$U_{\text{PMNS}} = \begin{pmatrix} 0.0008 - 0.7556 \ \ 0.0001 - 0.7497 \ \ 0.1886 - 0.999 \\ 0.0206 - 0.9998 \ 0.00178 - 0.9969 \ 0.00035 - 0.9815 \\ 0.00259 - 0.9967 \ 0.0002 - 0.9999 \ 0.00689 - 0.9964 \end{pmatrix}. \qquad (13)$$

A look at the above matrix reveals that several matrix elements, e.g. U_{e1} and U_{e3}, show no overlap with the corresponding one given in equation. This, therefore, immediately rules out FXGAR mass matrices for the IO case of neutrino masses.

These conclusions can also be checked further by graphs given in Fig. 1 wherein, the plots show the parameter space of two mixing angles obtained

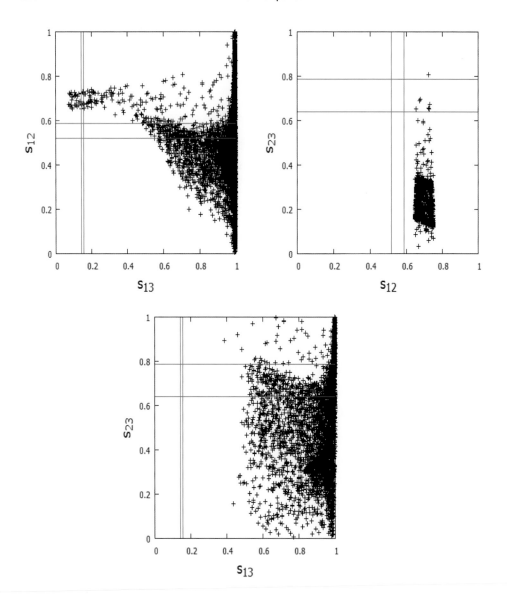

Fig. 1.   Plots showing the parameter space for any two mixing angles when the third angle is constrained by its 3σ range for IO of neutrino masses.

by constraining the third one by its 3σ range. The horizontal and vertical lines in these plots correspond to the 3σ ranges of the mixing angles on the corresponding axis. The region showing the overlap of the vertical and the horizontal lines represents the experimentally allowed 3σ ranges of the two mixing angles being considered. The parameter space obtained for these two mixing angles, after constraining the third one, is shown through points. One can clearly see that in all the three plots, the obtained parameter space of two mixing angles does not show any overlap with the experimentally allowed region. Therefore, FXGAR mass matrices are ruled out for the IO case of neutrino masses.

4.2. *Normal ordering of neutrino masses*

After ruling out the IO case, we now examine the compatibility of the matrices, given in Eq. (3) for the NO case. To this end, we have first constructed the PMNS matrix, which comes out to be as follows

$$U_{\text{PMNS}} = \begin{pmatrix} 0.801 - 0.845 \ 0.513 - 0.579 \ 0.142 - 0.155 \\ 0.224 - 0.460 \ 0.474 - 0.700 \ 0.630 - 0.776 \\ 0.283 - 0.528 \ 0.451 - 0.684 \ 0.612 - 0.760 \end{pmatrix}. \tag{14}$$

Interestingly, the above matrix shows good deal of compatibility with a recently constructed PMNS matrix by Garcia *et al.* [55], given in Eq. (12).

Similar to the plots given for the IO case, in Fig. 2 we present the plots showing the parameter space corresponding to any two mixing angles when the third one is constrained by its 3σ range for the NO case. Interestingly, one can see from these plots that the obtained parameter space shows significant overlap with the experimentally allowed 3σ range, shown by the rectangle formed at the intersection of vertical and horizontal lines, hence, confirming again that FXGAR mass matrices are viable for the NO case of neutrino masses.

After examining the viability, it becomes interesting to explore whether any constraints can be obtained on the lightest neutrino mass m_{ν_1} in this case. To this end, in Fig. 3 we have plotted the variation of the lightest neutrino mass with the mixing angles. For plotting these graphs, other two mixing angles have been constrained by their 3σ bounds. The parallel vertical lines in each plot show the 3σ range of the mixing angle being considered. A careful look at all these three graphs reveals that for the current experimental ranges of the three neutrino mixing angles, the lightest neutrino mass comes out to be nearly in the range 0.001-0.01 eV. Further, in order to explore the possibility of CP violation in the leptonic sector, in Fig. 4 we plot the dependence of Jarlskog's rephasing invariant parameter J on the three mixing angles, keeping the other two mixing angles constrained by their 3σ ranges. From these graphs, one finds J to lie in the range 0.0005–0.03, indicating likeliness of existence of CP violation in the leptonic sector.

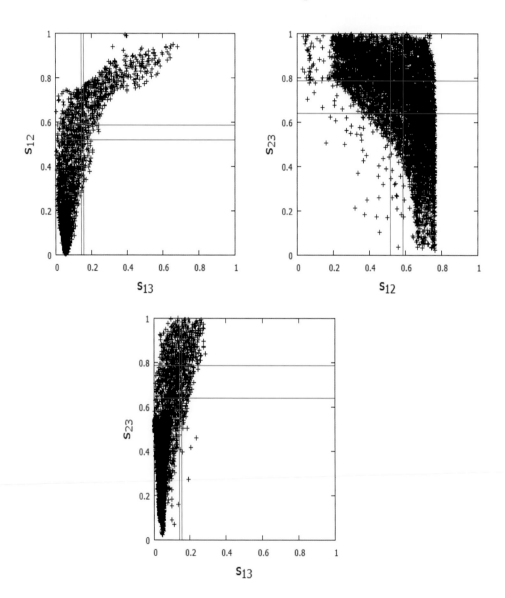

Fig. 2.   Plots showing the parameter space for any two mixing angles when the third angle is constrained by its 3σ range for NO of neutrino masses.

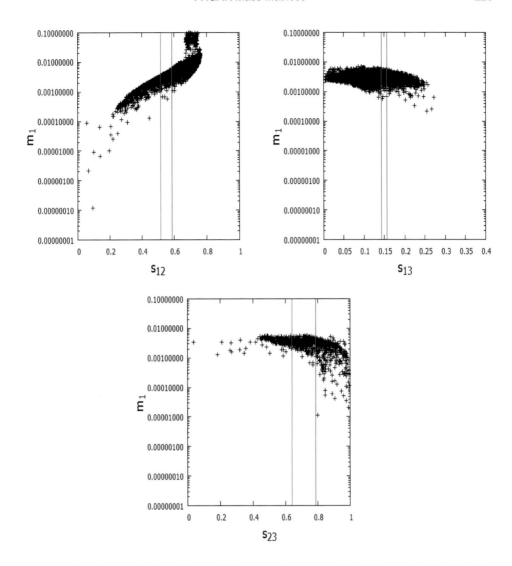

Fig. 3.   Plots showing the dependence of lightest neutrino mass on the mixing angles when the other two angles are constrained by their 3σ ranges.

5. Summary and Conclusions

FXGAR mass matrices are a particular class of texture 4 zero hermitian mass matrices, known to be quite successful in explaining the quark mixing data. Following the spirit of quark lepton symmetry, wherein quarks and leptons are treated on the same footing, assuming neutrinos to be Dirac particles, we have made an attempt to examine the compatibility of FXGAR mass

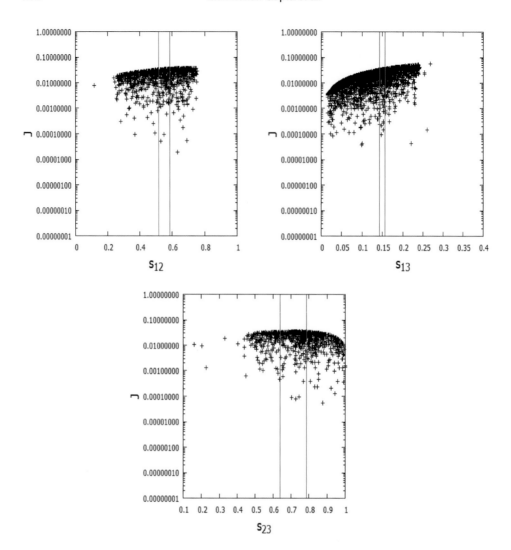

Fig. 4. Plots showing the variation of Jarlskog CP violating parameter with mixing angles when the other two angles are constrained by their 3σ ranges.

matrices with the latest data for the NO and IO cases of neutrino masses. Our analysis reveals that these mass matrices appear to be ruled out for the cases pertaining to IO of neutrino masses, while for NO case, these mass matrices appear to be a viable possibility. The implications of all three lepton mixing angles on the lightest neutrino mass as well as the Jarlskog's CP violating parameter in the leptonic sector for the NO case of neutrino masses have been investigated.

Acknowledgments

The authors MG and GA would like to record a special thanks to Prof. K.K. Phua for providing us the opportunity to visit Institute of Advanced Studies, NTU, Singapore on several occasions. Also, MG, GA, AB, MA and AV would like to thank Chairman, Department of Physics, Panjab University, for providing the facilities to work. KSC would like to thank Director, Department of Physics, Chandigarh University, for providing the facilities to work.

References

[1] N. Cabibbo, *Phys. Rev. Lett.* **10** (1963) 531.

[2] M. Kobayashi, T. Maskawa, *Prog. Theor. Phys.* **49** (1973) 652.

[3] P. A. Zyla *et al.* (Particle Data Group), *Prog. Theor. Exp. Phys.* **2020** (2020) 083C01.

[4] Guo-yuan Huang and Shun Zhou, *Phys. Rev. D* **103** (2021) 016010.

[5] N. Vinyoles, A. M. Serenelli, F. L. Villante, S. Basu, J. Bergstrm, M. C. Gonzalez-Garcia *et al.*, *Astrophys. J.* **835** (2017) 202.

[6] B. T. Cleveland*et al.*, *Astrophys. J.* **496** (1998) 505.

[7] F. Kaether, W. Hampel, G. Heusser, J. Kiko and T. Kirsten, *Phys. Lett. B* **685** (2010) 47.

[8] SAGE collaboration, J. N. Abdurashitov *et al.*, *Phys. Rev. C* **80** (2009) 015807.

[9] Super-Kamiokande collaboration, J. Hosaka *et al.*, *Phys. Rev. D* **73** (2006) 112001; Super-Kamiokande collaboration, J. Cravens *et al.*, *Phys. Rev. D* **78** (2008) 032002; Super-Kamiokande collaboration, K. Abe *et al.*, *Phys. Rev. D* **83** (2011) 052010. Y. Nakajima, "SuperKamiokande". Talk given at the *XXIX International Conference on Neutrino Physics and Astrophysics*, Chicago, USA, June 22–July 2, 2020.

[10] SNO collaboration, B. Aharmim *et al.*, *Phys. Rev. C* **88** (2013) 025501.

[11] Borexino collaboration, G. Bellini *et al.*, *Phys. Rev. Lett.* **107** (2011) 141302; *Phys. Rev. D* **82** (2010) 033006; *Nature* **512** (2014) 383.

[12] M. Honda, M. Sajjad Athar, T. Kajita, K. Kasahara and S. Midorikawa, *Phys. Rev. D* **92** (2015) 023004.

[13] IceCube collaboration, M. Aartsen *et al.*, *Phys. Rev. D* **91** (2015) 072004.

[14] Super-Kamiokande collaboration, K. Abe *et al.*, *Phys. Rev. D* **97** (2018) 072001.

[15] KamLAND collaboration, A. Gando *et al.*, *Phys. Rev. D* **88** (2013) 033001.

[16] T. Bezerra, New results from the double Chooz experiment. Talk given at the *XXIX International Conference on Neutrino Physics and Astrophysics*, Chicago, USA, June 22–July 2, 2020.

[17] Daya Bay collaboration, D. Adey *et al.*, *Phys. Rev. Lett.* **121** (2018) 241805.

[18] J. Yoo, RENO. Talk given at the *XXIX International Conference on Neutrino Physics and Astrophysics*, Chicago, USA, June 22–July 2, 2020.

[19] MINOS collaboration, P. Adamson *et al.*, *Phys. Rev. Lett.* **110** (2013) 251801.

[20] P. Dunne, Latest neutrino oscillation results from T2K. Talk given at the *XXIX International Conference on Neutrino Physics and Astrophysics*, Chicago, USA, June 22–July 2, 2020.

[21] A. Himmel, New oscillation results from the NOvA experiment. Talk given at the *XXIX International Conference on Neutrino Physics and Astrophysics*, Chicago, USA, June 22–July 2, 2020.

[22] J. C. Pati and A. Salam, *Phys. Rev. Lett.* **31** (1973) 661.

[23] J. Wess and B. Zumino, *Nucl. Phys. B* **70** (1974) 39.

[24] E. Farhi and L. Susskind, *Phys. Rev.* **74** (1981) 277 and references therein.

[25] M. Green and J. Schwarz, *Phys. Lett. B* **149** (1984) 117.

[26] H. Fritzsch and Z. Z. Xing, *Nucl. Phys. B* **556** (1999) 49 and references therein.

[27] H. Fritzsch and Z. Z. Xing, *Prog. Part. Nucl. Phys.* **45** (2000) 1.

[28] Z. Z. Xing and H. Zhang, *J. Phys. G* **30** (2004) 129 and references therein.

[29] M. Gupta and G. Ahuja, *Int. J. Mod. Phys. A* **26** (2011) 2973 and references therein.

[30] M. Gupta and G. Ahuja, *Int. J. Mod. Phys. A* **27** (2012) 1230033 and references therein.

[31] S. Weinberg, *Trans. N. Y. Acad. Sci., Series II* **38** (1977) 185.

[32] H. Fritzsch, *Phys. Lett. B* **70** (1977) 436.

[33] H. Fritzsch, *Phys. Lett. B* **73** (1978) 317.

[34] S. Kaundal, A. Bagai, G. Ahuja, M. Gupta, *Prog. Theor. Exp. Phys.* **2020**(1) (2020) 013B06.

[35] A. Bagai, S. Kaundal, G. Ahuja, M. Gupta, *Prog. Theor. Exp. Phys.* **2020**(9) (2020) 093B10.

[36] A. Bagai, A. Vashisht, N. Awasthi, arXiv:hep/ph-2110-05065.

[37] G. C. Branco *et al.*, *Phys. Rev. Lett.* **82** (1999) 683; *ibid. Phys. Lett. B* **477** (2000) 147, *ibid. Phys. Lett. B* **670** (2009) 340.

[38] P. S. Gill and M. Gupta, *J. Phys. G: Nuc. Part. Phys.* **21** (1995) 1; M. Randhawa, V. Bhatnagar, P. S. Gill and M. Gupta, *Phys. Rev. D* **60** (1999) 051301; M. Randhawa and M. Gupta, *Phys. Rev. D* **63** (2001) 097301; G. Ahuja, S. Kumar, M. Randhawa, M. Gupta and S. Dev, *Phys. Rev. D* **76** (2007) 013006; G. Ahuja, S. Sharma, P. Fakay and M. Gupta, *Mod. Phys. Lett. A* **30** (2015) 1530025.

[39] A. Yu. Smirnov, hep-ph/0604213.

[40] KamLAND-Zen collaboration, A. Gando *et al.*, *Phys. Rev. Lett.* **117** (2016) 082503.

[41] K. S. Babu and X. G. He, *Mod. Phys. Lett. A* **4** (1989) 61.

[42] J. T. Peltoniemi, D. Tommasini and J. W. F. Valle, *Phys. Lett. B* **298** (1993) 383.

[43] W. Wang, R. Wang, Z. L. Han and J. Z. Han, *Eur. Phys. J. C* **77** (2017) 889.

[44] C. Bonilla and J. W. F. Valle, *Phys. Lett. B* **762** (2016) 162.

[45] D. Borah and A. Dasgupta, *JHEP* **01** (2017) 072.

[46] S. Baek, *Phys. Lett. B* **805** (2020) 135415.

[47] J. Calle, D. Restrepo and A. Zapata, *Phys. Rev. D* **101** (2020) 035004.

[48] G. Ahuja, M. Gupta, M. Randhawa and R. Verma, *Phys. Rev. D* **79** (2009) 093006.

[49] X. W. Liu and S. Zhou, *Int. J. Mod. Phys. A* **28** (2013) 1350040.

[50] G. Ahuja and M. Gupta, *Int. J. Mod. Phys. A* **33** (2018) 1844032.

[51] L. M. Cebola, D. Emmanuel-Costa and R. G. Felipe, *Phys. Rev. D* **92** (2015) 025005.

[52] P. Fakay, S. Sharma, G. Ahuja and M. Gupta, *PTEP* **2014** (2014) 023B03.

[53] S. Sharma, P. Fakay, G. Ahuja and M. Gupta, arXiv:hep/ph-1402.0628.

[54] B. Pontecorvo, *Zh. Eksp. Teor. Fiz (JETP)* **33** (1957) 549; *ibid.* **34** (1958) 257; *ibid.* **53** (1967) 1717; Z. Maki, M. Nakagawa, S. Sakata, *Prog. Theor. Phys.* **28** (1962) 870.

[55] I. Esteban, M. C. Gonzalez-Garcia *et al.*, *J. High Energy Phys.* **2020** (2020) 178.

Lattice Nonlinear Schrödinger Model: History, Applications and Open Problems

Kun Hao[*,†] and Vladimir Korepin[†,‡]

[*]*Institute of Modern Physics, Northwest University, Xi'an 710127, China*
[†]*C.N. Yang Institute for Theoretical Physics, Stony Brook University, NY 11794, USA*
[*]*haoke72@163.com*
[‡]*korepin@gmail.com*

The lattice nonlinear Schrödinger model has applications in many fields. In this paper, we briefly review the basic concepts of the model. Special emphasis is given to the algebraic content of quantum integrable models. Quantum lattice nonlinear Schrödinger is equivalent to XXX spin chain with negative spin. It can be solved by the quantum inverse scattering method (algebraic Bethe ansatz). For the spin $s = -1$ case, XXX spin chain appears as an effective theory of Quantum Chromodynamics. The Yang–Yang equation describes quantum thermodynamics of lattice nonlinear Schrödinger model. We also provide the conformal field theory description and entanglement entropy time evolution of the model. Symmetry, correlations, and other applications are mentioned.

1. Formulation of the Model

The continuous nonlinear Schrödinger equation has many different names: Lieb–Liniger model, one-dimensional Bose gas with δ interaction, Tonks–Girardeau gas (infinite coupling constant).

The Hamiltonian of the model is:

$$H = \int dx(\partial_x \psi^\dagger \partial_x \psi + \kappa \psi^\dagger \psi^\dagger \psi \psi), \qquad \kappa \geq 0 \qquad (1)$$

Here ψ is the Bose field (dynamical variable). In the classical case, it is a complex function. κ is the coupling constant. The corresponding equation of motion,

$$i\partial_t \psi = -\partial_x^2 \psi + 2\kappa \psi^\dagger \psi \psi \tag{2}$$

is called the nonlinear Schrödinger equation.

The Poisson brackets of the fields ψ, $\psi^\dagger$ give

$$\{\psi(x), \psi^\dagger(y)\} = i\delta(x - y) \tag{3}$$

This model is the classical limit of the quantum lattice nonlinear Schrödinger (NLS) equation. It is integrable both in classical and quantum cases. It has infinitely many conservation laws.

In the quantum case, ψ and $\psi^\dagger$ become operators. The Poisson brackets become commutation relations,

$$[\psi(x), \psi^\dagger(y)] = \delta(x - y) \tag{4}$$

The corresponding equations of motion can be represented in the following Lax from,[a]

$$\partial_t L_n = M_{n+1}(\lambda) L_n(\lambda) - L_n(\lambda) M_n(\lambda). \tag{5}$$

The integer n is the discrete space variable, $x = n\Delta$, and Δ is the lattice step.

The Lax pair $L_n(\lambda)$ and $M_n(\lambda)$ are two-dimensional matrices. Their entries are expressed in terms of dynamical variables ψ of the lattice system. They also depend on spectral parameter λ. In quantum case, it is sufficient to have L operator and R matrix in order to apply quantum inverse scattering method [3].

We shall mainly focus on the quantum case [2]. The quantum monodromy matrix and transfer matrix are defined by

$$T_0(\lambda) = L_N(\lambda) \cdots L_2(\lambda) L_1(\lambda), \quad \tau(\lambda) = tr_0 \, T_0(\lambda) \tag{6}$$

Here N is the total lattice sites. The subscript 0 denotes the auxiliary space. The Lax representation follows from Yang–Baxter equation,

$$R(\lambda, \mu) \left(L_n(\lambda) \otimes L_n(\mu) \right) = \left(L_n(\mu) \otimes L_n(\lambda) \right) R(\lambda, \mu) \tag{7}$$

---

[a]This is true for both classical case and quantum case. The proofs are given in Refs. [1] and [2], respectively.

C. N. Yang found the following $R(\lambda, \mu)$ in Ref. [4]. It is the solution of the Yang–Baxter equation:

$$R(\lambda, \mu) = R(\lambda - \mu) = i\mathbb{1} + (\mu - \lambda)\Pi = \begin{pmatrix} \mu - \lambda + i & 0 & 0 & 0 \\ 0 & i & \mu - \lambda & 0 \\ 0 & \mu - \lambda & i & 0 \\ 0 & 0 & 0 & \mu - \lambda + i \end{pmatrix} \quad (8)$$

Here λ (also μ) is the spectral parameter. i is the crossing parameter. This is also the R matrix which describes the XXX Heisenberg spin-$\frac{1}{2}$ chain [5]. Here Π is the permutation operator. The R matrix has the following properties

Initial condition: $R(0) = i\mathbb{1}$,   (9)

Unitarity: $R(\lambda - \mu)R_{21}(\mu - \lambda) = (i - \lambda + \mu)(i + \lambda - \mu)\mathbb{1}$,   (10)

Projection: $R(i) = i(\mathbb{1} - \Pi)$.   (11)

Note that for $\lambda - \mu = i$ the $R(\lambda, \mu) = i(\mathbb{1} - \Pi)$ is degenerate. It is the 1D antisymmetric projector.

1.1. *Quantum inverse scattering method formulation*

P. P. Kulish and L. D. Faddeev formulated algebraic Bethe Ansatz [6] and embedded the quantum NLS model into quantum inverse scattering method(QISM). An approximate L operator [7] (up to the first order of Δ) on a dense lattice is:

$$L_n(\lambda) = \begin{pmatrix} 1 - \frac{i\lambda\Delta}{2} & -i\sqrt{\kappa}\chi_n^{\dagger} \\ i\sqrt{\kappa}\chi_n & 1 + \frac{i\lambda\Delta}{2} \end{pmatrix} + O(\Delta^2), \quad (12)$$

The integer n is a discrete space variable $x = n\Delta$, and Δ is the step of the lattice. χ_n is the quantum Bose field:

$$[\chi_n, \chi_m^{\dagger}] = \Delta\,\delta_{nm} \quad (13)$$

Here λ is the spectral parameter and κ is the coupling constant. $\chi = \psi\Delta$ in the limit $\Delta \to 0$.

The exact lattice Lax operator[b] (all orders in Δ) for both the classical case and the quantum case was constructed by A. Izergin and V. Korepin in

[b]In classical case, the commutation relations become $\{\chi_m, \chi_n^{\dagger}\} = i\delta_{mn}\Delta$.

Ref. [2], see also Ref. [8]

$$L_j(\lambda) = \begin{pmatrix} 1 - \frac{i\lambda\Delta}{2} + \frac{\kappa}{2}\chi_j^\dagger\chi_j & -i\sqrt{\kappa}\chi_j^\dagger\varrho_j \\ i\sqrt{\kappa}\varrho_j\chi_j & 1 + \frac{i\lambda\Delta}{2} + \frac{\kappa}{2}\chi_j^\dagger\chi_j \end{pmatrix}, \tag{14}$$

$$[\chi_j, \chi_l^\dagger] = \Delta\delta_{j,l} \quad \text{and} \quad \varrho_j = (1 + \frac{\kappa}{4}\chi_j^\dagger\chi_j)^{\frac{1}{2}}. \tag{15}$$

Here $\kappa > 0$, and $\Delta > 0$. It has the same R matrix (8).

We use these notations to define the classical Hamiltonian of NLS

$$H_c = \frac{i}{12\kappa}\left(\frac{d}{d\lambda^{-1}}\right)^3 \ln\left[(1 + \lambda/v)^{-N}\,\tau(\lambda)\right] + \text{complex conjugate} \tag{16}$$

The explicit expression shows that this Hamiltonian describes the interaction of five nearest neighbors on the lattice. In the continuous limit $(\chi_n = \psi_n\Delta; \psi_{n+1} - \psi_n = O(\Delta); \Delta \to 0, N \to \infty$ but $N\Delta = \text{const})$ Eq. (16) goes to the correct Hamiltonian of the continuous model (1) and L_n operator (14) turns into correct continuous L_n operator, see Ref. [9].

The quantum version Hamiltonian of NLS has the form:

$$H_q = \left(\frac{i}{12\kappa}\left(\frac{d}{d\lambda^{-1}}\right)^3 + \frac{i\kappa}{6}\frac{d}{d\lambda^{-1}}\right)\ln\left[(1 + \lambda/v)^{-N}\,\tau(\lambda)\right] + \text{hermitian conjugate} \tag{17}$$

The model can be solved by QISM [6]. The pseudo-vacuum Ω is annihilated by lattice Bose fields $\chi_n\Omega = 0$.

1.2. *Equivalence of quantum NLS and XXX chain with negative spin*

The quantum lattice NLS model can be interpreted as the XXX Heisenberg chain with negative spin s (not necessarily half-integral [2]). In the classical continuous case, they are gauge equivalent [10].

We can rewrite the L operator as XXX Heisenberg chain [11]:

$$L_j^{XXX} = -\sigma^z L_j = i\lambda + S_j^k \otimes \sigma^k \tag{18}$$

$$S_j^+ = -i\sqrt{\kappa}\chi_j^\dagger\varrho_j, \quad S_j^- = i\sqrt{\kappa}\varrho_j\chi_j, \quad S_j^z = (1 + \frac{\kappa}{2}\chi_j^\dagger\chi_j). \tag{19}$$

The σ are Pauli matrices. The S_j form a representation of $SU(2)$ algebra with negative spin

$$s = -\frac{2}{\kappa \Delta} \tag{20}$$

Our main example will be $s = -1$ (for $\Delta = 2$ and $\kappa = 1$).

1.3. *Inversion*

Let us be reminded of the definition of the monodromy matrix and transfer matrix:

$$T_0(\lambda) \equiv L_N(\lambda) \ldots L_1(\lambda) = \begin{pmatrix} A(\lambda) & B(\lambda) \\ C(\lambda) & D(\lambda) \end{pmatrix}_0, \qquad \tau(\lambda) = tr_0\, T_0(\lambda). \tag{21}$$

Subscript N is the length of the lattice. Based on (7), the commutation relations among the elements of the monodromy matrix in auxiliary space 0 are given by the R matrix

$$R(\lambda, \mu)\left(T_0(\lambda) \bigotimes T_0(\mu)\right) = \left(T_0(\mu) \bigotimes T_0(\lambda)\right) R(\lambda, \mu) \tag{22}$$

The inverse of quantum monodromy matrix is called antipode:

$$T(\lambda)^{-1} = d_q^{-N}(\lambda)\sigma^y T^t(\lambda + i)\sigma^y \tag{23}$$

$$d_q(\lambda) = \Delta^2(\lambda - v)(\lambda - v + i)/4. \tag{24}$$

Here $v = -2i/\Delta$. This is a deformation of the Cramer's rule [12]. The difference is a shift of the spectral parameter by i.

The denominator is the quantum determinant:

$$\det_q T(\lambda) = A(\lambda)D(\lambda + i) - B(\lambda)C(\lambda + i) = d_q^N(\lambda) \tag{25}$$

The quantum determinant [2] is a complex valued function, but not a quantum operator. More precisely, it is proportional to the identity operator in the quantum space, with the coefficient being a $\mathbb{C}$-number.

The quantum determinant describes the center of Yang–Baxter algebra in quantum groups. It is also used in the theory of correlation functions and form factors.

2. Heisenberg XXX Chain with Negative Spin and Quantum Nonlinear Schrödinger Model

The XXX chain with negative spin can be considered as discretization of quantum lattice nonlinear Schrödinger equation (NLS) [2,13,14]. It has applications in high energy physics.

The holomorphic QCD describing the deep-inelastic scattering (L. Lipatov and G. Korchemsky) at small Bjorken[c] x is dual to the XXX spin $s = -1$ chain. It is solvable by Bethe Ansatz.

At high energy, the scattering amplitudes are described by the exchange of gluons dressed by virtual gluon loops: so-called Reggeized gluons. In the limit of large number of colors N_c (with fixed $g^2 N_c$, where g is the QCD coupling), the corresponding Feynman diagrams have the topology of the cylinder. The Hamiltonian describing the interactions of reggeized gluons reduces to the sum of terms describing the pairwise near-neighbor interactions: the XXX spin chain [15,16].

The scattering of a lepton on a hadron is a sum of Feynman diagrams. In leading logarithmic approximation (LLA), ladder diagrams dominate. Quarks exchange gluons. The Hamiltonian of NLS can describe interactions of the gluons [17]. The small Bjorken x is

$$x = \frac{Q^2}{2pq}, \quad Q^2 = -q^2$$

The p is the hadron momentum and q is the photon momentum.

A hint for derivation: Vladimir Gribov and his school established that "ladder" Feynman diagrams mainly contribute at large momentum. Balitski, Fadin, Kuraev and Lipatov (BFKL) wrote a linear integral equation for the sum of the ladder diagrams. In 1993, Lipatov made the Fourier transform of the kernel of the BFKL [18–21] equation. Originally it is in two dimensions, from transverse momenta to position space. The result is the holomorphic anti-holomorphic factorization and the two-dimensional conformal (Möbius) symmetry of the kernel in this position representation. The Fourier transformation is non-trivial, singular integrals. The result allows to

---

[c]See http://www.scholarpedia.org/article/Bjorken_scaling

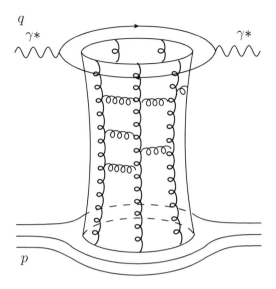

q

$\gamma*$

$\gamma*$

p

Fig. 1.   Feynman diagram describing deep inelastic scattering at small Bjorken x. The virtual photon $\gamma*$ emitted by the scattered lepton (not shown) splits into a virtual quark–antiquark pair. The reggeized gluons are exchanged between the virtual quark–antiquark pair and the hadron.

find the solution of the BFKL equation for the general case of non-vanishing momentum transfer in terms of conformal 3-point-functions. Lipatov shows that the Hamiltonian of his spin chain commutes with the non-local integrals derived from the monodromy of the product of the $sl(2)$ L matrices in the Jordan–Schwinger form [22,23].

The algebraic Bethe Ansatz (quantum inverse scattering method) [14,24] can be applied to the solution of equations for wave functions of compound states which describe L reggeized gluons[d] in the multicolor QCD in a generalized leading logarithmic approximation.

In the following, we will first briefly introduce the connection of QCD in the LLA limit to the XXX spin chain for spin $s = -1$, and the relation between XXX spin chain with negative spin and quantum nonlinear Schrödinger model. These relations enable us to further investigate the thermodynamics of the models.

---

[d]We remark that in the whole section for the holomorphic QCD, L is the number of reggeized gluons (particles). This follows from the literature in high energy physics. The notation is slightly different from other integrable models. For detailed explanation, see Refs. [17] and [25].

In quantum case, the Hamiltonian of NLS was proposed by Tarasov, Takhtadzhyan and Faddeev. It describes the interaction of nearest neighbors:

$$\mathcal{H} = \sum_{k=1}^{L} H_{k,k+1} \tag{26}$$

The definition of the chain is based on the existence of a fundamental matrix $R_{jk}^{(s,s)}(\lambda)$ which obeys the Yang–Baxter equation

$$R_{jk}^{(s,s)}(\lambda) = f(s,\lambda)\frac{\Gamma(i\lambda - 2s)\Gamma(i\lambda + 2s + 1)}{\Gamma(i\lambda - J_{jk})\Gamma(i\lambda + J_{jk} + 1)}. \tag{27}$$

Here $f(s,\lambda)$ is a complex valued function (it normalizes the R matrix). The λ is called spectral parameter. The superscript (s,s) means that both the auxiliary space and the quantum space have spin s. The operator J_{jk} is defined in the space $V \otimes V$ as a solution of the operator equation,

$$J_{jk}(J_{jk} + 1) = 2\vec{S}_j \otimes \vec{S}_k + 2s(s + 1). \tag{28}$$

L. Lipatov and G. Korchemsky considered deep inelastic scattering (DIS) at small Bjorken x. The holomorphic multicolor QCD Hamiltonian describes the nearest neighbor interactions of L particles (reggeized gluons) with periodic boundary conditions $H_{L,L+1} = H_{L,1}$. The local Hamiltonians are given by the equivalent representations

$$\begin{aligned} H_{j,k} &= -P_j^{-1}\ln(z_{jk})P_j - P_k^{-1}\ln(z_{jk})P_k - \ln(P_jP_k) - 2\gamma_E \\ &= -2\ln(z_{jk}) - (z_{jk})\ln(P_jP_k)(z_{jk})^{-1} - 2\gamma_E, \end{aligned} \tag{29}$$

where $P_j = i\partial/\partial z_j = i\partial_j$, $z_{jk} = z_j - z_k$, and γ_E is the Euler constant. It describes the deep inelastic scattering of an electron on a nucleon.

Lipatov used holomorphic representation of $SU(2)$

$$S_k^+ = z_k^2\partial_k - 2sz_k, \qquad S_k^- = -\partial_k, \qquad S_k^z = z_k\partial_k - s, \tag{30}$$

with $k = 1, \cdots, L$. He mapped DIS (29) to XXX spin chain with $s = 0$ and -1.

The Hamiltonian of the XXX model with spin $s = 0$ describes the interaction of nearest neighbors, see (26)

$$H_{jk} = \frac{-1}{i}\frac{d}{d\lambda}\ln R_{jk}^{(s=0)}(\lambda)\Big|_{\lambda=0} = -\psi(-J_{jk}) - \psi(J_{jk} + 1) + 2\psi(1). \tag{31}$$

Here $\psi(x) = d \ln \Gamma(x)/dx$, and $\psi(1) = -\gamma_E$ (γ_E is the Euler constant). The operator J_{jk} is a solution of (28) when $s = 0$,

$$J_{jk}(J_{jk} + 1) = -(z_j - z_k)^2 \partial_j \partial_k. \tag{32}$$

Take $j = k + 1$ and substitute into (26). This is a description of DIS in QCD by $s = 0$ spin chain.

After a similarity transformation, the spin $s = 0$ model can be mapped to $s = -1$ model. The latter can be easily solved by algebraic Bethe Ansatz method. Thus the high-energy asymptotics in multicolor QCD is exactly solvable, and it has the same eigenvalues with that of XXX spin $s = -1$ chain. As we mentioned, the XXX chain with negative spin is equivalent to the discretization of nonlinear Schrödinger equation (NLS).

3. Thermodynamics Bethe Ansatz for Quantum Lattice NLS

The thermodynamics Bethe Ansatz method is based on C. N. Yang and C. P. Yang's work [26], in which they studied the thermodynamics of continuous NLS.

Let us consider the correspondence between the Bethe equations of the two models. The Bethe roots λ_k of the quantum lattice NLS model satisfy the following Bethe equations. When we take coupling constant $\kappa = 1$, and $\Delta = 2$, the quantum lattice NLS Bethe equations become that of the XXX spin chain model with spin $s = -1$

$$\left(\frac{i(\frac{-2}{\kappa \Delta})\kappa + \lambda_k}{i(\frac{-2}{\kappa \Delta})\kappa - \lambda_k} \right)^L = \prod_{j \neq k}^{N} \frac{\lambda_k - \lambda_j + i\kappa}{\lambda_k - \lambda_j - i\kappa} \quad \xrightarrow[s=-1]{\kappa=1, \Delta=2}$$

$$(-1)^L \left(\frac{\lambda_k - i}{\lambda_k + i} \right)^L = \prod_{j \neq k}^{N} \frac{\lambda_k - \lambda_j + i}{\lambda_k - \lambda_j - i}. \qquad k = 1, \cdots, N. \tag{33}$$

This means that quantum lattice NLS model describes a more general XXX spin chain model with negative spin $s = -2/\kappa\Delta$. The holomorphic QCD is its special case with spin $s = -1$, $\Delta = 2$ and coupling constant $\kappa = 1$.

These are periodic boundary conditions. The energy spectrum is:

$$E = \sum_{j=1}^{N} \frac{-2}{\lambda_j^2 + 1} \tag{34}$$

In order to analyze the Bethe roots distribution for negative spin $s = -1$, we first take the logarithm of Bethe equations (33) for quantum lattice NLS,

$$2\pi n_k = \sum_{j=1}^{N} \theta(\lambda_k - \lambda_j) + L\,\theta(\lambda_k), \tag{35}$$

where

$$\theta(\lambda) = -\theta(-\lambda) = i\ln\left(\frac{i+\lambda}{i-\lambda}\right); \quad -\pi < \theta(\lambda) < \pi, \quad \operatorname{Im}\lambda = 0. \tag{36}$$

This function is monotonically increasing,

$$\theta'(\lambda - \mu) = K(\lambda, \mu) = \frac{2}{1 + (\lambda - \mu)^2}, \quad K(\lambda) = K(\lambda, 0). \tag{37}$$

Here n_k are different integer (or half integer) numbers. They satisfy the Pauli principle [27]. All λ_k are also different.

3.1. Analysis of Bethe equations

Theorem 1. *If solutions of Bethe equations (Eq. (33) for $s = -1$) exist then they are real numbers.*

$$\left(\frac{\lambda_k - i}{\lambda_k + i}\right)^L = \prod_{j=1,\ j\neq k}^{N} \frac{\lambda_k - \lambda_j + i}{\lambda_k - \lambda_j - i}, \quad k = 1, \cdots, N. \tag{38}$$

Proof. Let us use the following properties of the Bethe equations:

LHS: $\left|\dfrac{\lambda - i}{\lambda + i}\right| \le 1$, when $\operatorname{Im}\lambda \ge 0$; $\left|\dfrac{\lambda - i}{\lambda + i}\right| \ge 1$, when $\operatorname{Im}\lambda \le 0$; (39)

RHS: $\left|\dfrac{\lambda + i}{\lambda - i}\right| \ge 1$, when $\operatorname{Im}\lambda \ge 0$; $\left|\dfrac{\lambda + i}{\lambda - i}\right| \le 1$, when $\operatorname{Im}\lambda \le 0$; (40)

If we denote the one with maximal imaginary part as $\lambda_{\max} \in \{\lambda_j\}$, then

$$\operatorname{Im}\lambda_{\max} \ge \operatorname{Im}\lambda_j, \quad j = 1, \cdots, N. \tag{41}$$

Taking the modulus of both sides of the equation for $\lambda_j = \lambda_{\max}$, make use of the estimate for the right-hand side (40) and obtain

$$\left|\frac{\lambda_{\max} - i}{\lambda_{\max} + i}\right|^L = \left|\prod_{j=1}^{N} \frac{\lambda_{\max} - \lambda_j + i}{\lambda_{\max} - \lambda_j - i}\right| \ge 1. \tag{42}$$

Due to LHS, this results in:

$$\text{Im}\,\lambda_j \leq \text{Im}\,\lambda_{\max} \leq 0, \quad j = 1, \cdots, N. \tag{43}$$

Similarly, one has

$$0 \leq \text{Im}\,\lambda_{\min} \leq \text{Im}\,\lambda_j, \quad j = 1, \cdots, N. \tag{44}$$

Thus the only remaining possibility is $\text{Im}\,\lambda_j = 0, j = 1, \cdots, N$. Theorem 1 is proved. No bond states exist in this repulsive case.

3.2. Yang–Yang action is convex

Theorem 2. *The solutions of the logarithmic form Bethe equations (35) exist.*

Proof. The proof procedure [14] is based on the fact that Eq. (35) can be obtained from a variational principle. The Yang–Yang action was introduced by C. N. Yang and C. P. Yang in Ref. [26]:

$$S = L \sum_{k=1}^{N} \theta_1(\lambda_k) + \frac{1}{2} \sum_{k,j}^{N} \theta_1(\lambda_k - \lambda_j) - 2\pi \sum_{k=1}^{N} n_k \lambda_k, \tag{45}$$

where $\theta_1(\lambda) = \int_0^\lambda \theta(\mu)d\mu$. Logarithmic Bethe equations are the extremum (minimum) conditions for S: $\partial S / \partial \lambda_j = 0$.

To prove this, it is sufficient to establish that the matrix of second derivatives is positive definite.

$$\frac{\partial^2 S}{\partial \lambda_j \partial \lambda_l} = \delta_{jl}[L\,K(\lambda_j) + \sum_{m=1}^{N} K(\lambda_j, \lambda_m)] - K(\lambda_j, \lambda_l) \tag{46}$$

Consider some real vector v_j. The quadratic form is positive:

$$\sum_{j,l} \frac{\partial^2 S}{\partial \lambda_j \partial \lambda_l} v_j v_l = \sum_{j=1}^{N} L\,K(\lambda_j)v_j^2 + \sum_{j>l} K(\lambda_j, \lambda_l)(v_j - v_l)^2 \geq 0. \tag{47}$$

The $K(\lambda_j)$ are positive. The action is convex: it has unique minimum. It defines the solution of Bethe equation and is unique (in the logarithmic form). Theorem 2 is proved.

We remark that the determinant of the matrix (46) gives the square of the norm of the Bethe wave function. The corresponding formula in the book

[14] is (0.3) in the Introduction section of Chapter X. So the wave function does not vanish.

3.3. *The thermodynamic limit at zero temperature*

Considering the more general case, for positive κ, all λ_j have to be different: Pauli principle in the momentum space is valid [27]. In the limit $L \rightarrow \infty$ and $N \rightarrow \infty$, the λ_j are condensed into Fermi sphere $[-q, q]$. The particle distribution function $\rho_p(\lambda_j) = \frac{1}{L(\lambda_{j+1} - \lambda_j)}$ satisfies:

$$2\pi\rho_p(\lambda) = \int_{-q}^{q} K(\lambda, \mu)\rho_p(\mu)d\mu + K(\lambda) \tag{48}$$

$$K(\lambda, \mu) = \frac{2\kappa}{\kappa^2 + (\lambda - \mu)^2}, \quad K(\lambda) = K(\lambda, 0). \tag{49}$$

In this case we have to require that

$$\int_{-q}^{q} \rho_p(\lambda)d\lambda = D = \frac{N}{L} \tag{50}$$

For $\kappa = 0$ Fermi sphere collapse: the ground state is Bose–Einstein condensate.

Open problem is to analyze the integral equation in the limit of $\kappa \rightarrow 0$ and describe singularities. $K(\lambda, \mu) \rightarrow 2\pi\delta(\lambda - \mu)$: will the integral cancel the left-hand side?

3.4. *Collapse of the Fermi sphere in the continuous case*

In the continuous case (Bose gas with δ interaction) of nonlinear Schrödinger the integral equation is:

$$2\pi\rho_p(\lambda) = \int_{-q}^{q} K(\lambda, \mu)\rho_p(\mu)d\mu + 1. \tag{51}$$

This is Lieb–Liniger equation. In the limit $\kappa \rightarrow 0$, $K(\lambda, \mu) \rightarrow 2\pi\delta(\lambda - \mu)$. The integral cancels on the left-hand side. The limit was studied by S. Prolhac [28], G. Lang [29], and C. Tracy & H. Widom [30]. The decomposition is in $\sqrt{\kappa}$ and $\log \kappa$. Coefficients are objects of number theory. In the lattice case the limit is an open problem.

3.5. *Energy of elementary excitation*

Return to the lattice case: special case XXX with spin $s = -1$. Considering the grand canonical ensemble, the energy spectrum becomes

$$E_h = \sum_{j=1}^{N} (\frac{-2}{\lambda_j^2 + 1} - h) \tag{52}$$

The h is the chemical potential. In thermodynamic limit, the energy of elementary excitation $\varepsilon(\lambda)$ satisfies the linear integral equation

$$\varepsilon(\lambda) - \frac{1}{2\pi} \int_{-q}^{+q} K(\lambda, \mu)\varepsilon(\mu)d\mu = \frac{-2}{\lambda^2 + 1} - h \equiv \varepsilon_0(\lambda), \tag{53}$$

$$\varepsilon(q) = \varepsilon(-q) = 0. \tag{54}$$

The elementary excitation has a topological charge: it does not fit into periodic boundary conditions. We have to change periodic boundary conditions to anti-periodic boundary conditions.

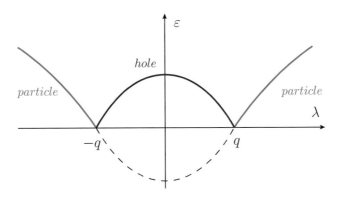

Fig. 2.   The energy of the elementary excitation as a function of λ. For $-q < \lambda < q$ elementary excitation is a hole, but it is the particle for other values of λ.

The energy of the elementary excitation is a function of λ. For $-q < \lambda < q$ elementary excitation is a hole, but it is the particle for other values of λ.

In the infinite volume limit, any energy level is a scattering state of several elementary excitations with different momenta.

3.6. *Momentum of the elementary excitation*

The momentum of the particle $k(\lambda_p)$ is

$$k(\lambda_p) = p_0(\lambda_p) + \int_{-q}^{q} \Theta(\lambda_p - \mu)\rho_p(\mu)d\mu, \quad \Theta(\lambda) = p_0(\lambda) = i\ln\left(\frac{i+\lambda}{i-\lambda}\right). \quad (55)$$

The momentum $k_h(\lambda_h)$ of elementary hole excitation is

$$k_h(\lambda_h) = -p_0(\lambda_h) - \int_{-q}^{q} \Theta(\lambda_h - \mu)\rho_p(\mu)d\mu. \quad (56)$$

where $-q < \lambda_h < q$. At zero temperature, all the observables are described by a linear integral equation.

3.7. *Scattering matrix*

The NLS model has infinitely many conservation laws. This imposes strong limitations on scattering. For example, there is no reflection during scattering of particle and a hole. The scattering matrix of two elementary excitations is a transition coefficient:

$$S = \exp\{-i\phi(\lambda_p, \lambda_h)\}, \quad (57)$$

here the scattering phase satisfies the integral equation:

$$\phi(\lambda_p, \lambda_h) - \frac{1}{2\pi}\int_{-q}^{+q} K(\lambda_p, v)\phi(v, \lambda - \lambda_h)dv = \Theta(\lambda_p - \lambda_h). \quad (58)$$

$$\Theta(\lambda) = -\Theta(-\lambda) = i\ln\left(\frac{i\kappa + \lambda}{i\kappa - \lambda}\right); \quad -\pi < \Theta(\lambda) < \pi, \quad \text{Im}\,\lambda = 0 \quad (59)$$

Many body scattering matrix is a product of pairwise scattering matrices. This can be used as a definition of complete integrability in many body quantum mechanics.

3.8. *Quantum thermodynamics*

Following the works [5,14], one derives the Yang–Yang equation for current model by using variation and the steepest descent method. The Yang–Yang equation describes quantum thermodynamics of NLS:

$$\varepsilon(\lambda) = \frac{-2}{\lambda^2+1} - h - \frac{T}{2\pi}\int_{-\infty}^{+\infty} K(\lambda, \mu)\ln(1 + e^{-\varepsilon(\mu)/T})d\mu, \quad (60)$$

with the following properties:

$$\frac{\rho_h(\lambda)}{\rho_p(\lambda)} = e^{\varepsilon(\lambda)/T}, \qquad D = \frac{N}{L} = \int_{-\infty}^{\infty} \rho_p(\lambda)d\lambda. \tag{61}$$

The $\varepsilon(\lambda)$ is the energy of the stable excitation. $\rho_p(\lambda)$ and $\rho_h(\lambda)$ are densities of particles and holes, respectively. An open problem is to consider the analytical solution in limiting case: $\kappa \to 0$, then $K(\lambda, \mu) \to 2\pi\delta(\lambda - \mu)$.

Based on Yang–Yang equation, one can further derive some physical quantities like free energy, pressure and entropy. The free energy is:

$$\mathcal{F} = Nh - \frac{LT}{2\pi} \int_{-\infty}^{+\infty} K(\mu) \ln(1 + \exp(-\varepsilon(\mu)/T))d\mu. \tag{62}$$

The pressure is:

$$\mathcal{P} = -\left(\frac{\partial \mathcal{F}}{\partial L}\right)_T = \frac{T}{2\pi} \int_{-\infty}^{+\infty} K(\mu) \ln(1 + e^{-\varepsilon(\mu)/T})d\mu. \tag{63}$$

Thermal entropy is:

$$S = -\frac{\partial \mathcal{F}}{\partial T} = \frac{L}{2\pi} \int_{-\infty}^{+\infty} K(\mu) \left[\ln(1 + e^{-\varepsilon(\mu)/T}) + \frac{\varepsilon(\mu)}{T(e^{\varepsilon(\mu)/T} + 1)} \right] d\mu. \tag{64}$$

The thermodynamics of NLS was realized experimentally in quantum optics. It was built in optical lattice by N. J. van Druten [31].

4. Entanglement Entropy Evolution

As we mentioned, the XXX spin chain with negative spin can be mapped to the quantum lattice nonlinear Schrödinger's model. After evaluation of finite size corrections, we conclude the Virasoro algebra (describing effective CFT) has the central charge equal to one $c = 1$. Based on the CFT description, we find that the time evolution of entanglement entropy after local quench is logarithmic

$$\mathcal{S}_A(t) = \frac{c}{3} \log(mt), \qquad \frac{1}{m} < t < \frac{1}{xm} \tag{65}$$

given by the CFT description of NLS. m is the mass of the proton. After critical time $t_c = 1/(mx)$, the entanglement saturates. This result was published in January 2022, the first issue of *Physical Review D* [25]. The logarithmic dependence on time was first reported by P. Calabrese and J. Cardy [32].

4.1. *CFT description of entanglement entropy dynamics*

Let us consider the entropy of a block of spins (an internal of infinite system). At positive temperature, thermal fluctuations dominate. A theory of classical shock waves[e] shows that after local quench the entropy is a linear function of time:

$$S(t) = \left(\frac{2\pi c}{3}T\right)t \tag{66}$$

The coefficient can be found in Ref. [33]; See for example their formula (80). T is the temperature. The shock wave changes the density of the entropy.

At zero temperature the entropy is some function f of time $S(t) = f(t)$. Conformal mapping shows that for positive $T > 0$

$$S_T(t) = f\left(\frac{v}{\pi T} \sinh \frac{\pi T}{v}(x + vt)\right), \tag{67}$$

where v is velocity. Considering the limit of large time, the entanglement entropy at time t is

$$S_T(t) = f\left(\exp[\pi T(t - t_0)]\right). \tag{68}$$

Now we have two expressions for the entropy of positive temperature

$$S_T(t) = \frac{2\pi c}{3}Tt = f\left(\exp[\pi T(t - t_0)]\right). \tag{69}$$

This means the function is given by

$$f(t) = \frac{2c}{3}\log t + \text{const.} \tag{70}$$

Each end of the block contributes equally, so for the local quench

$$S(t) = \frac{c}{3}\log t + \text{const.} \tag{71}$$

This agrees with the entanglement entropy evolution after local quench (with one point of defect) calculated in [32]. D. Kharzeev and E. Levin compared our formula to gluon distribution function measured on HERA.

---

[e]Time evolution of classical entropy and classical shock waves was developed by Rudolf Clausius. It shows that after local quench [explosion] a shock wave emerges.

4.2. *Entanglement entropy of a block of spins*

At zero temperature the ground state $|gs\rangle$ is unique. The entropy of the ground state is zero. Let us consider a block of x sequential lattice cites. We interpret the rest of the lattice as an environment. We trace away the environment: this gives us the density matrix of the block $\rho = tr_E\left(|gs\rangle\langle gs|\right)$. The von Neumann entropy of the block is a complicated function of x, but for large x, it scales logarithmically

$$S_{vN} = -tr(\rho \log \rho) \rightarrow \frac{1}{3}\log(x) \quad \text{as} \quad x \rightarrow \infty. \tag{72}$$

It is similar to the continuous case of NLS [34].

We remark that the logarithm is typical for gapless models, but not universal. In spin chains based on information theory, the entropy scales as a square root of x [35,36]. In fireball model of nucleus proposed by Rolf Hagedorn in 1960,[f] the entropy increases exponentially.

4.3. *Rényi entropy*

The Rényi entropy is defined as

$$S_R = \frac{\log\left(tr\rho^\alpha\right)}{1-\alpha}, \quad \alpha > 0. \tag{73}$$

Here α is a new parameter. It is a real number. For NLS, the Rényi entropy also scales logarithmically with the size of a block

$$S \rightarrow \frac{(1+\alpha^{-1})\log x}{6} \tag{74}$$

as in XX spin chain [37].

In some spin chains, the Rényi entropy is not an analytical function of α: it scales with x differently for different α [38].[g]

5. Symmetry and Correlations

5.1. *Correlation functions*

In continuous NLS case, the correlation was evaluated in the papers by A. Klümper, A. Foerster [39] and Herman Boos, Frank Göhmann [40].

---

[f]See https://en.wikipedia.org/wiki/Hagedorn_temperature
[g]This is Stokes phenomenon https://en.wikipedia.org/wiki/Stokes_phenomenon

First at zero temperature, time independent in the infinite volume, correlation functions in XXX spin-$\frac{1}{2}$ chain can be expressed as polynomials (with rational coefficients) of the values of Riemann zeta function with odd arguments.[h] See Ref. [41] by H. Boos, V. Korepin.

More works are available on correlation functions for exactly solvable spin chains by Andreas Klümper, Dominic Nawrath [42], H. Boos, F. Göhmann, A. Klümper, J. Suzuki [43–45], and T. Miwa, F. Smirnov [46].

Open problems are the following: Can we calculate correlation functions in XXX chain with negative spin? Can we describe correlation functions of the XXX with negative spin by number theory?

5.2. *Extra symmetry in the infinite volume*

At spin-$\frac{1}{2}$ the XXX chain gains an additional symmetry in the thermodynamic limit. It is the YANGIAN SYMMETRY (infinite dimensional quantum group). See Ref. [47] for a general introduction, as well as Hubbard model [48]. The Yangian symmetry also appears in supersymmetric Yang-Mills 4D: L. Corcoran, F. Loebbert, J. Miczajka, M. Staudacher [49], and N. Beisert, A. Garus, M. Rosso [50]. R. Kirschner suggests Yangian as a tool for the investigation of integrability of QCD [51].

An open problem is that does an additional symmetry arise in XXX with negative spin in the limit of infinitely long lattice?

6. Discretization
6.1. *Another integrable discretization*

M. Ablowitz and J. Ladik [52] constructed a different integrable discretization of nonlinear Schrödinger. The L operator is different.

$$\frac{i}{2}\frac{\partial}{\partial t}\psi(n,t) = (1 + 4\psi(n,t)\psi^\dagger(n,t))(\psi(n+1,t) + \psi(n-1,t)), \qquad (75)$$

$$-\frac{i}{2}\frac{\partial}{\partial t}\psi^\dagger(n,t) = (1 + 4\psi(n,t)\psi^\dagger(n,t))(\psi^\dagger(n+1,t) + \psi^\dagger(n-1,t)). \qquad (76)$$

---

[h]The values of Riemann zeta function with odd arguments are celebrated in number theory, see https://en.wikipedia.org/wiki/Particular_values_of_the_Riemann_zeta_function They are conjectured to be transcendental numbers, algebraically independent over the field of rational numbers, see Wikipedia: Apery's theorem.

Relation between different discretizations of nonlinear Schrödinger equation is interesting even in the classical case. Tim Hoffmann [53] proved that in the classical case the L operator of Ablowitz–Ladik version is gauge equivalent to the Izergin-Korepin version of NLS.

Classical Ablowitz–Ladik has important applications, see Ref. [54]. It describes space, time and temperature dependent correlation function in a spin chain.

6.2. *Double discrete version*

In the classical case, Tim Hoffmann constructed an integrable double discrete version of nonlinear Schrödinger and related it to geometry: Discrete Hashimoto Surfaces and a Doubly Discrete Smoke-Ring Flow.[i] See the chapter [55] of the book *Discrete Differential Geometry*.

The field was developed by Alexander Bobenko and Yuri Suris. They wrote the book: *Discrete Differential Geometry, Integrable Structure* [56]. It has applications to architecture. There is no Hamiltonian formulation, just recursion relations. The quantization of the double discrete nonlinear Schrödinger remains an open problem.

7. Conclusion

Quantum integrability was discovered by H. Bethe in 1931 [57] "Zur Theorie der Metalle. I. Eigenwerte und Eigenfunktionen der linearen Atomkette". Zeitschrift für Physik. 71 (March 1931): 205–226. Since that time, a lot of progress has been made. Still, new developments appear. One recent progress is that the algebraic Bethe Ansatz is on a quantum computer now. The Qiskit program, which inputs algebraic Bethe ansatz in a quantum computer, was recently written by Alejandro Sopena, Max Hunter Gordon, Diego García-Martín, Germán Sierra, and Esperanza López [58]. There are a lot of open problems for young people.

---

[i]The modern way is to describe them by tropical geometry https://en.wikipedia.org/wiki/Tropical_geometry

Acknowledgments

The work of KH was supported by the National Natural Science Foundation of China (Grant Nos. 11805152, 12047502 and 11947301), the Natural Science Basic Research Program of Shaanxi Province Grant Nos. 2021JCW-19 and 2019JQ-107, and Shaanxi Key Laboratory for Theoretical Physics Frontiers in China. VK was supported by the SUNY center for QIS at Long Island project number CSP181035. VK would like to thank Prof. K.K. Phua for the hospitality during his visit to Nanyang Technological University in Singapore.

References

[1] E. K. Sklyanin, Quantum version of the method of inverse scattering problem, *J. Sov. Math.* **19** (1982) 1546–1596.

[2] V. E. Korepin and A. G. Izergin, A lattice model connected with a nonlinear Schrödinger equation, *Dokl. Akad. Nauk SSSR* **259** (1981) 76–79.

[3] L. Faddeev and L. Takhtajan, *Hamiltonian Methods in the Theory of Solitons*, Springer Series in Soviet Mathematics (Springer, 2007).

[4] C. N. Yang, Some exact results for the many-body problem in one dimension with repulsive delta-function interaction, *Phys. Rev. Lett.* **19** (1967) 1312–1315.

[5] C. N. Yang, s matrix for the one-dimensional n-body problem with repulsive or attractive δ-function interaction, *Phys. Rev.* **168** (1968) 1920–1923.

[6] L. D. Faddeev, *How the Algebraic Bethe Ansatz Works for Integrable Models*, (World Scientific, 2016) pp. 370–439.

[7] E. K. Sklyanin and L. D. Faddeev, *Quantum-mechanical Approach to Completely Integrable Field Theory Models*, (World Scientific, 2016) pp. 290–292.

[8] A. G. Izergin and V. E. Korepin, Lattice versions of quantum field theory models in two dimensions, *Nucl. Phys. B* **205**(3) (1982) 401–413.

[9] V. E. Zakharov and A. B. Shabat, Exact theory of two-dimensional self-focusing and one-dimensional self-modulation of waves in nonlinear media, *Sov. J. Experim. Theoret. Phys.* **34** (1972) 62.

[10] V. E. Zakharov and L. A. Takhtadzhyan, Equivalence of the nonlinear Schrödinger equation and the equation of a Heisenberg ferromagnet, *Theoret. Math. Phys.* **38** (1979) 17–23.

[11] W. Heisenberg, Zur theorie des ferromagnetismus, *Zeitschrift für Physik* **49** (1928) 619–636.

[12] Z. Gong, M. Aldeen and L. Elsner, A note on a generalized Cramer's rule, *Linear Algebra Appl.* **340**(1) (2002) 253–254.

[13] V. E. Korepin and A. G. Izergin, Lattice model connected with nonlinear Schrodinger equation, *Sov. Phys. Doklady* **26** (1981) 653–654.

[14] V. E. Korepin, N. M. Bogoliubov and A. G. Izergin, *Quantum Inverse Scattering Method and Correlation Functions*, Cambridge Monographs on Mathematical Physics (Cambridge University Press, 1993).

[15] L. N. Lipatov, High energy asymptotics of multi-color qcd and exactly solvable lattice models, *Pisma Zh. Eksp. Teor. Fiz.* **59** (1994) 571–574.

[16] L. N. Lipatov, High energy asymptotics of multi-color qcd and exactly solvable lattice models, *JETP Lett.* **59** (1994) 596–599.

[17] K. Hao, D. Kharzeev and V. E. Korepin, Bethe ansatz for XXX chain with negative spin, *Int. J. Mod. Phys. A* **34**(31) (2019) 1950197.

[18] L. N Lipatov, Reggeization of the vector meson and the vacuum singularity in nonabelian gauge theories, *Sov. J. Nucl. Phys.* **23**(3) (1976) 338.

[19] E. A. Kuraev, L. N. Lipatov and V. S. Fadin, Multi-reggeon processes in yang-milles theory, *Sov. Phys. JETP* **44**(3) (1976) 443.

[20] E. A. Kuraev, L. N. Lipatov and V. S. Fadin, The pomeranchuk singularity in nonabelian gauge theories, *Sov. Phys. JETP* **45** (1977) 199.

[21] Y. Y. Balitskii and L. N. Lipatov, Pomeranchuk singularity in quantum chromodynamics, *Sov. J. Nucl. Phys. (Engl. Transl.); (United States)* **28**(6) (1978).

[22] L. N. Lipatov, High energy asymptotics of perturbative QCD, *Acta Phys. Pol. B* **26** (1995) 1815.

[23] L. N. Lipatov, Small-x physics in perturbative QCD, *Phys. Rep.* **286**(3) (1997) 131–198.

[24] L. D. Faddeev, Algebraic aspects of the Bethe Ansatz, *Int. J. Mod. Phys. A* **10**(13) (1995) 1845–1878.

[25] K. Zhang, K. Hao, D. Kharzeev and V. Korepin, Entanglement entropy production in deep inelastic scattering, *Phys. Rev. D* **105** (2022) 014002.

[26] C. N. Yang and C. P. Yang, Thermodynamics of a one-dimensional system of bosons with repulsive delta-function interaction, *J. Math. Phys.* **10**(7) (1969) 1115–1122.

[27] A. G. Izergin and V. E. Korepin, Pauli principle for one-dimensional bosons and the algebraic Bethe ansatz, *Lett. Math. Phys.* **6** (1982) 283–288.

[28] S. Prolhac, Ground state energy of the delta-bose and fermi gas at weak coupling from double extrapolation, *J. Phys. A: Mathematical and Theoretical* **50**(14) (2017) 144001.

[29] G. Lang, Conjectures about the structure of strong and weak coupling expansions of a few ground-state observables in the Lieb-Liniger and Yang-Gaudin models, *SciPost Phys.* **7** (2019) 55.

[30] C. A. Tracy and H. Widom, On the ground state energy of the delta-function Fermi gas II: Further asymptotics, *Geometric Methods in Physics XXXV*, eds. P. Kielanowski, A. Odzijewicz and E. Previato (Springer International Publishing, Cham, 2018), pp. 209–220.

[31] A. H. van Amerongen, J. J. P. van Es, P. Wicke, K. V. Kheruntsyan and N. J. van Druten, Yang–Yang thermodynamics on an atom chip, *Phys. Rev. Lett.* **100** (2008) 090402.

[32] P. Calabrese and J. Cardy, Entanglement and correlation functions following a local quench: A conformal field theory approach, *J. Stat. Mech.: Theory and Experiment*, **2007**(10) (2007) P10004.

[33] P. Calabrese and J. Cardy, Entanglement entropy and conformal field theory, *J. Phys. A: Mathematical and Theoretical*, **42**(50) (2009) 504005.

[34] V. E. Korepin, Universality of entropy scaling in one dimensional gapless models, *Phys. Rev. Lett.* **92** (2004) 096402.

[35] O. Salberger and V. Korepin, *Fredkin Spin Chain*, (World Scientific, 2018), pp. 439–458.

[36] R. Movassagh and P. W. Shor, Supercritical entanglement in local systems: Counterexample to the area law for quantum matter, *Proc. Nat. Acad. Sci.*, **113**(47) (2016) 13278–13282.

[37] B.-Q. Jin and V. E. Korepin, Quantum spin chain, Toeplitz determinants and the Fisher-Hartwig conjecture, *J. Stat. Phys.* **116** (2004) 79–95.

[38] F. Sugino and V. Korepin, Rényi entropy of highly entangled spin chains, *Int. J. Mod. Phys. B* **32**(28) (2018) 1850306.

[39] O. I. Pâţu, A. Klümper and A. Foerster, Universality and quantum criticality of the one-dimensional spinor Bose gas, *Phys. Rev. Lett.* **120** (2018) 243402.

[40] H. Boos and F. Göhmann, On the physical part of the factorized correlation functions of the XXZ chain, *J. Phys. A: Mathematical and Theoretical* **42**(31) (2009) 315001.

[41] H. E. Boos and V. E. Korepin, Quantum spin chains and Riemann zeta function with odd arguments, *J. Phys. A: Mathematical and General* **34**(26) (2001) 5311–5316.

[42] A. Klümper, D. Nawrath and J. Suzuki, Correlation functions of the integrable isotropic spin-1 chain: Algebraic expressions for arbitrary temperature, *J. Stat. Mech.: Theory and Experiment*, **2013**(08) (2013) P08009.

[43] H. E. Boos, F. Göhmann, A. Klümper and J. Suzuki, Factorization of multiple integrals representing the density matrix of a finite segment of the Heisenberg spin chain, *J. Stat. Mech.: Theory and Experiment* **2006**(04) (2006) P04001–P04001.

[44] H. E. Boos, F. Göhmann, A. Klümper and J. Suzuki, Factorization of the finite temperature correlation functions of the XXZ chain in a magnetic field, *J. Phys.*

A: Mathematical and Theoretical **40**(35) (2007) 10699–10727.

[45] H. E. Boos, J. Damerau, F. Göhmann, A. Klümper, J. Suzuki and A. Weiße, Short-distance thermal correlations in the XXZ chain, *J. Stat. Mech.: Theory and Experiment* **2008**(08) (2008) P08010.

[46] T. Miwa and F. Smirnov, New exact results on density matrix for XXX spin chain, *Lett. Math. Phys.* **109** (2019) 675–698.

[47] D. Bernard, An introduction to Yangian symmetries, *Int. J. Mod. Phys. B* **07**(20n21) (1993) 3517–3530.

[48] D. B. Uglov and V. E. Korepin, The Yangian symmetry of the hubbard model, *Phys. Lett. A* **190**(3) (1994) 238–242.

[49] L. Corcoran, F. Loebbert, J. Miczajka and M. Staudacher, Minkowski box from Yangian bootstrap, *J. H. Energy Phys.* **2021** (2021) 160.

[50] N. Beisert, A. Garus and M. Rosso, Yangian symmetry for the action of planar $\mathcal{N} = 4$ super Yang-Mills and $\mathcal{N} = 6$ super Chern-Simons theories, *Phys. Rev. D* **98** (2018) 046006.

[51] R. Kirschner, QCD at high energies and Yangian symmetry, *Universe* **4**(11) (2018).

[52] M. J. Ablowitz and J. F. Ladik, Nonlinear differential-difference equations and Fourier analysis, *J. Math. Phys.* **17**(6) (1976) 1011–1018.

[53] T. Hoffmann, On the equivalence of the discrete nonlinear Schrödinger equation and the discrete isotropic Heisenberg magnet, *Phys. Lett. A* **265**(1) (2000) 62–67.

[54] A. R. Its, A. G. Izergin, V. E. Korepin and N. A. Slavnov, Temperature correlations of quantum spins, *Phys. Rev. Lett.* **70** (1993) 1704–1706.

[55] T. Hoffmann, *Discrete Hashimoto Surfaces and a Doubly Discrete Smoke-Ring Flow*, (Birkhäuser Basel, Basel, 2008) pp. 95–115.

[56] A. I. Bobenko, J. M. Sullivan, P. Schröder and G. M. Ziegler (eds.), *Discrete Differential Geometry* (Birkhäuser Basel, Basel, 2008).

[57] H. Bethe, Zur theorie der metalle, *Zeitschrift für Physik* **71** (1931) 205–226.

[58] A. Sopena, M. H. Gordon, D. García-Martín, G. Sierra and E. López, Algebraic Bethe circuits, arXiv preprint arXiv:2202.04673, 2022.

Quantum Walk as a Simulator of Quantum Radar under Noisy Environments

Tsung-Wei Huang and Chih-Yu Chen

Undergraduate Program in Intelligent Computing and Big Data,
Quantum Information Center, Chung Yuan Christian University, Taoyuan, Taiwan

Ching-Ray Chang

Department of Physics and Quantum Information Center,
Chung Yuan Christian University, Taoyuan, Taiwan

Quantum radar is a recent emerging technology due to intensive studies of quantum computers, quantum computing and quantum information. Not only because of its excellent error-probabilities and high resolution, but it also has potential applications in long-range quantum sensing for the detection of stealth aircraft and short-range detectors for autonomous vehicles. In this paper, in addition to a brief review of the basic ideas of quantum information theory and how to build a quantum radar scheme, we also use quantum random walks to simulate the interaction of photons with noisy environments. Our results show that only dephasing interaction can preserve entangled states for entangled photon pairs from quantum radar if there are various possible forms of action in the environment. Other decoherent forms of action will easily destroy the quantum entangled state and affect the detection range and function of the quantum radar.

1. Introduction

Radar (Radio Detection and Ranging, RADAR) [1–3] is the abbreviation and transliteration of radio detection and ranging [Fig. 1(a)]. It emits electromagnetic energy into space in a directional manner. By receiving the radio waves reflected by objects in the sky, the direction, height and speed of the object can be calculated, and the shape of the object can be detected. With the rapid development of stealth technology and electronic jamming technology, stealth fighters have become the main attack force on the battlefield, and traditional radars have become useless. If there are stealth missiles in the future, it will be even more difficult to guard against them on the battle-field. Because of this, how to effectively detect stealth aircraft and projectile weapons has become a battleground for strategists.

Quantum radar [4] is an advanced version of traditional radar, and it is also a modern technology product developed for the detection of stealth aircraft [Fig. 1(b)]. As conventional radar, quantum radar also has a trans-mit antenna and a receive antenna. The transmit antenna shoots a set of entangled photons, one into the air and the other directly at the receive an-tenna. Since this group of photons are entangled with each other, when the quantum wave in the air hits the invisible flying object, the invisible flying object in the air can be easily detected due to the entanglement property of photons. Another advantage of quantum radar is interference immunity. This is mainly because the characteristics of the photons will be changed during quantum measurement, and it can be found whether the photon has been disturbed in the air by regularly detecting the characteristics of the photons, which can effectively prevent being deceived. With the maturity of

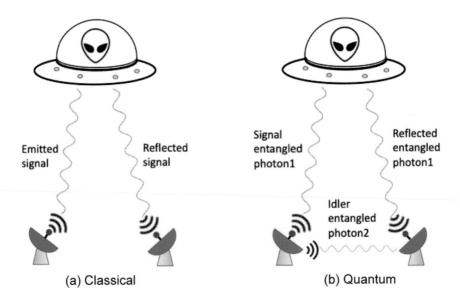

(a) Classical   (b) Quantum

Fig. 1. (a) The classical radar emits microwave to detect unknown flying objects from the detection of reflected microwave. (b) Quantum radar emits an entangled photon pair to both receiver antenna and the unknown flying objects. The reflected photon can be measured together with the idler photon and the quantum entangled properties can greatly enhance the detection sensitivity.

quantum radar technologies, almost all air stealth targets can be monitored and tracked in the future, which is the nemesis of expensive stealth fighters. Quantum radar has the characteristics of distinguishing invisible and stealth targets. It is the clairvoyance of future technological battlefields, and various stealth objects can be seen at any time.

Although quantum radar has a quantum advantage in detecting stealth fighters, it is also well known that quantum systems are very delicate and fragile, vulnerable to environmental influences. Therefore, how to analyze and understand the quantum state change of the entangled quantum pairs emitted by quantum radar after exposure to the atmospheric environment has become an important issue. The previous analysis is usually based on hybrid method which combines the classical and quantum approaches together. Using photons as quantum system while using classical approaches to simulate the environments [5], the full quantum description of entangled photons and environments is still lacking. This paper will use quantum random walks (QRWs) [6] to simulate the changes in the environment of particles and entangled photon pairs, and then hope to increase the robustness of quantum radar. QRWs are simple frameworks for simulating quantum phenomena, and there are two main types of QRWs, continuous-time quantum walks (CTQWs) and discrete-time quantum walks (DTQWs). In CTQW, walking is directly defined in a Hilbert space called position space H_p, while DTQW introduces an additional coin space H_c to define the direction of walking. Although both QRWs have their own advantages, DTQW is easier to implement in current quantum computers. We use quantum walks here to simulate the effects of quantum radar's entangled photons colliding with particles in the environment. We use CNOT gates and CS (controlled-S) gates to represent decoherence and dephasing interactions, respectively, and then analyze how photons change their quantum states during these two conditions.

2. Basic Theories for Quantum Radar

Quantum radar is a remote sensing technology using quantum entanglement and quantum illumination technology to detect quantum results at the receiver. When the electromagnetic wave hits the target and returns, many of the initial entanglement will be lost due to quantum decoherent collisions during propagation, but the system can pick out only the photons that were originally sent by the radar, completely filtering out any other sources.

(i) The entanglement of quantum states

The unique characters of quantum technology are superposition and entanglement. Consider a photon scattered by a target particle, Eq. (1) is an example of the superposition state of a single photon

$$|\psi_i\rangle = \frac{1}{\sqrt{2}}(|H_i\rangle + |V_i\rangle) \tag{1}$$

An entangled state for two photons reads

$$|\psi\rangle = \frac{1}{\sqrt{2}}(|H_1 H_2\rangle + |V_1 V_2\rangle) \tag{2}$$

where $|H\rangle$ ($|V\rangle$) represents the horizontal (vertical) polarization with respect to the scattering plane. State $|\psi\rangle$ cannot be further decomposed into a tensor product of two individual states. That is, $|\psi\rangle \neq |\psi_1\rangle \otimes |\psi_2\rangle$, where $|\psi_i\rangle$ is defined as in Eq. (1).

Entanglement plays an important role in quantum information. The connection between entanglement and entropy has been studied in many literatures, e.g. Refs. [7] and [8]. The article of Pechanski and Seki deals with entanglement entropy of scattering particles [9]. The study of quantum information is somewhat beyond the scope of this article. One can also find more complete contents in books or review articles for relative topics, see also, Refs. [10–12]. We merely focus on the behavior of entanglement phenomenally.

(ii) Quantum illumination

The current mainstream theory for implementing the quantum radar is via quantum illumination (QI). The idea of QI was first proposed by Lloyd and collaborators at MIT in 2008 [13], for object detection. In the QI mechanism, a source sender creates an entangled photon pair, called idler photon and signal photon, respectively. These two photons are initially in a quantum entangled state. The concept of QI is sending the signal photon out to interact with the remote object. In the meantime, the idler photon remains at the place where the source sender is.

When operating the detection, an observer (the source sender) keeps sending the signal photons and collecting the photons reflected by the remote object. The reflecting (signal) photons recombine with the idler system. The process above will repeat many times. The advantage of QI is it applies the quantum character of signal and idler photons to enhance the detection threshold. Classically the signal photons are easily affected by thermal fluctuation or other environmental factors. Now since the initial signal-idler state is quantum-correlated, we can impose some unique quantum fingerprint in the very beginning. That allows people to distinguish the reflecting signal even though the background photons (uncorrelated) are also received by the detector. People define error-probabilities (Pr) in order to judge the quality of a radar. A comparison is done between single photon (SP) and QI in different regimes [13].

Table 1. As shown in Ref. [13], the comparison of error-probabilities (Pr) between SP and QI in good/bad-regime. There is no significant advantage of QI in the "good regime". However, the factor M is the exponent enhanced successive rate of QI in the "bad regime". One can find more details and definition of parameters in Ref. [13].

Regime \ Type	Single photon (SP)	Quantum illumination (QI)
Good-regime	$Pr(e)_{SP} \leq \frac{1}{2}\exp(-N\kappa)$	$Pr(e)_{QI} \leq \frac{1}{2}\exp(-N\kappa)$
Bad-regime	$Pr(e)_{SP} \leq \frac{1}{2}\exp(\frac{-N\kappa^2}{8N_B})$	$Pr(e)_{QI} \leq \frac{1}{2}\exp(\frac{-N\kappa^2 M}{8N_B})$

Right after Lloyd's pioneer work, Tan *et al.* demonstrated the error-probability exponents of QI could be up to 6dB over the optimum reception coherent-state system [15]. In 2015, Barzanjeh *et al.* theoretically showed better error-probability exponents of QI compared to coherent-state sensors in the microwave regime [16]. Later in 2017, Zhuang *et al.* published two papers which discussed QI can enhance the signals in a noisy environment or for Rayleigh-fading targets [17,18]. Ranjith Nair and Mile Gu used quantum information theory to estimate the fundamental limit of QI in their 2020 paper [19]. Recently, Zhuang and Shapiro further proposed QI range is a quantum pulse-compression radar that benefits from the entanglement between a high time-bandwidth product transmitted signal pulse and a high time-bandwidth product retained idler pulse in 2021 [20]. For more comprehensive review of quantum illumination or quantum radar, one can find some review articles such as Refs. [4, 21] and the references therein.

(iii) Simulation of photons in a noisy environment

All practical facilities must work in the environment, and therefore must be affected by the environment, and even interact with the environment. The normal working environment of the quantum radar system will not be in the laboratory, but in the atmospheric environment full of various particles. Therefore, it is important to maintain good function of quantum radar in the noisy environment. For microwave systems, the environmental impact is discussed in Ref. [22], and it is concluded that operating at cryogenic temperatures can reduce the environmental impact. In Lloyd's paper [13], the authors further pointed out the advantages of quantum (illumination) radar in noisy environments. At the same time, in the microwave field, quantum radar is not only significantly better than traditional radar, but even better than quantum lidar due to advantage of long wavelength.

In 2015, Lanzagorta theoretically analyzed the possibility for low-brightness quantum radar. He showed that quantum radar significantly improves in many aspects such as imaging or anti-noise in low visibility earth atmosphere [23]. Saleh compared parameters between traditional and quantum radars in error probability upper bound (ε), signal-to-noise ratio (SNR) in his paper [24]. He also concluded that quantum radar has better resolution. The analysis above is a hybrid analysis applying both quantum and classical approaches, it treats photons as quantum system while using classical approaches to simulate the background, see also Ref. [5].

Despite Ref. [23] having predicted the performance of quantum radar is better compared to the traditional one, however, the behavior of photon state decoherent is not clear in the semi-classical approach. Here we use quantum random walk (QRW) [6] to simulate the decoherent mechanism of entangled photons with the particles in environments. Conventionally, quantum states $|0\rangle$ and $|1\rangle$ represent the position of an object in real space in the QRW method. In our model, we use the number of qubits to represent different quantum states of emitting photons in the environments. For example, four entangled states $|00\rangle, |01\rangle, |10\rangle, |11\rangle$ represent $|H\rangle, \alpha|H\rangle + i\beta|V\rangle, \alpha|H\rangle - i\beta|V\rangle, |V\rangle$ state, respectively. The $|H\rangle$ and $|V\rangle$ are basis defined conventionally in quantum optics [24] or in Eq. (2) and the constraint $|\alpha|^2 + |\beta|^2 = 1$ holds. The probabilistic coin in QRW determines whether or not the quantum state varies owing to the interaction between photons and particles in the environment. The photon-particle interactions are simulated via control gates in quantum computer.

(iv) Quantum Random Walk

In classical random walk (CRW), one uses a coin to determine the direction of an object. People apply a similar idea into a quantum system. We define Hilbert space $H = H_c \otimes H_p$ where H_c and H_p is the coin space and position space, respectively. The coin space H_c for one-dimensional QRW is a two-dimensional Hilbert space. The coin operator C is a 2×2 matrix, in fact, C can be a unitary operator of any kind. In contrast to the classical system, the measurement critically influences the results in a quantum system. Mathematically, we use shift operators to link the coin space and the position space by Eq. (3). In a quantum circuit, we use a controlled gate to read the coin state so that the shift operator changes the position based on the read-out value. The controlled gate is generally a CNOT gate since it can be viewed as the object moves to $+1$ or -1, analogous to a classical coin in CRW.

The gate algorithm of QRW for the Discrete Time Quantum Walk (DTQW) is shown in Fig. 2, where the coin operator is C and shifts conditional operator is S. For general DTQW, the coin operator is any unitary matrix and conditional shift operator is controlled by the state of coin space H_I,

$$S = |0\rangle_{cc}\langle 0| \otimes \sum_x |x-1\rangle\langle x| + |1\rangle_{cc}\langle 1| \otimes \sum_x |x+1\rangle\langle x|. \qquad (3)$$

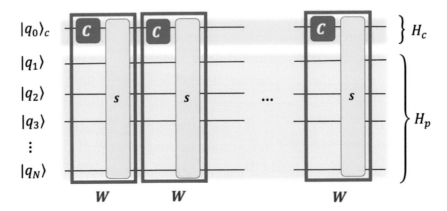

Fig. 2.   Discrete Time Quantum Walk algorithm. Each step is composed of a unitary coin operator C and followed by a unitary conditional shift operator S.

3. Quantum Random Walk Results

We are here trying to simulate a process where a beam of photons is injected into an environment with thermal noise. We assumed that the thermal noise is a randomly distributed perturbation without specific directions, see Fig. 3. For simplicity, we also neglect the level differences of the atmosphere as well as the atmospheric absorption. The model simply treats the background as white noise. Even this toy model is simple, however, it still demonstrates the basic features of quantum radar within a noisy environment. In particular, it is a good model to simulate a small scale (1km or less) application.

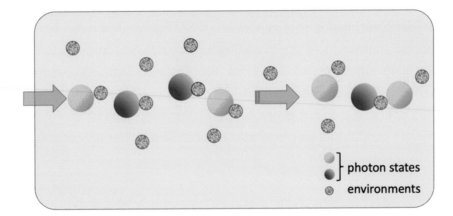

Fig. 3.   A photon enters an environment with thermal noise. The coin of the QRW determines the probability of a photon changing its state or not, e.g. from $|00\rangle$ to $|00\rangle$ or remaining in $|00\rangle$.

In this circumstance, an initially entangled photon pair can be assumed as

$$\frac{1}{\sqrt{2}}(|00\rangle_1 \otimes |00\rangle_2 + |11\rangle_1 \otimes |11\rangle_2) \tag{4}$$

The indices 1 and 2 in Eq. (4) means the first and second photons, respectively. Thus state $|0010\rangle$ that stands for photon 1 is $|00\rangle$ and photon 2 is $|10\rangle$ in abbreviation. The probability of changing the photon state, for example $|00\rangle \rightarrow |01\rangle$, is 50% for each event when a photon collides with environmental particles from thermal agitations. We apply a CNOT gate to simulate such change-of-state mechanism. The interaction influences on the states changing can be viewed as a strongly decoherent environment. The QRW results under strong decoherence interaction are shown in Fig. 4. With an initially entangled state as Eq. (4), all possible states show up even under a single collision [Fig. 4(a)] and no state presents convex probability distribution. Because the probability distribution of all states is quite uniformly distributed, we can view the outcome of each walk (interaction) is irrelevant with the previous state. This implies that the correlation between the sequence does not exist and it is a Markovian process. Thus, the strong decoherence interaction of QRW destructs the entanglement immediately.

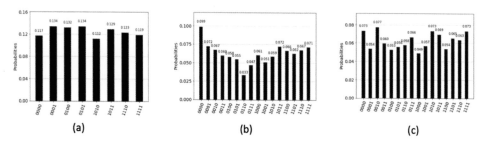

Fig. 4.   QRW using CNOT gates, after (a) one walk (collision) (b) 5 walks (collisions) (c) 14 walks (collisions). The distribution of states after strong decoherence interaction shows no obvious correlation even after a few steps of interactions.

Now if we replace CNOT gates by CS gates, this means the environment affects the phase of the incoming photon with correlated $|H\rangle$ and $|V\rangle$ states. In this case, the final state demonstrates quite interesting behaviors after a few steps. As shown in Fig. 5, the final state of a photon tends to converge to a mixed state $\frac{1}{\sqrt{2}}(|0101\rangle + |1010\rangle)$. In fact, the final state is quite stable after

7 walks (not shown here) and behaves as that evolving into an equilibrium state. We see that the initial state $\frac{1}{\sqrt{2}}(|0000\rangle + |1111\rangle) = \frac{1}{\sqrt{2}}(|HH\rangle + |VV\rangle)$. It is not hard to verify that $\frac{1}{\sqrt{2}}(|0101\rangle + |1010\rangle)$ represents the same entangled state if the definitions $|01\rangle| \equiv \alpha|H\rangle + i\beta|V\rangle)$ and $|10\rangle \equiv \alpha|H\rangle - i\beta|V\rangle)$ are used. Therefore, the character of entanglement is still preserved in contrast to the decoherent outcomes in Fig. 4.

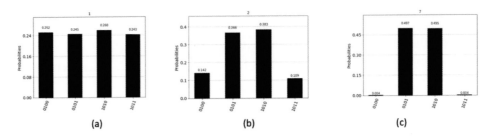

Fig. 5.   QRW using CS gates, after (a) one walk (collision) (b) 5 walks (collisions) (c) 14 walks (collisions). The distribution of states tends to shift into another stable final quantum state.

The property preservation of photon entangled states is crucial as it is the fundamental reason for enhanced detection of remote objects in quantum radar schemes. Scattering of nanoparticles in the atmosphere disrupts quantum properties and weakens the impact on quantum radar, due to quantum mechanisms whereby different types of interactions with the environment can change the state of photons. The present results from QRWs at least demonstrate how to maintain quantum entanglement in complex environmental interactions and will be a key technology for the success of quantum radar. Here the CS gate only varies the phase of $|H\rangle$ and $|V\rangle$ in the initial states but still remains in the entanglement.

4. Conclusion and Discussion

To understand the operation mechanism of a quantum radar system, it is also worthwhile to estimate the energy scale of a photon and noise. The operation range of the QI prototype quantum radar in the microwave regime is 5 to 12 GHz ($3.3 \times 10^{-24} - 7.9 \times 10^{-24}$ J). For an optical photon, the frequency is roughly 500 THz (3.3×10^{-19} J) and the thermal energy (k_BT) is about 4.1×10^{-21} J at room temperature. Obviously, thermal noise for an optical photon is a negligible effect but for the microwave it is quite significant. Thus many experiments in the optical domain can be performed at room temperature. However, to maintain an entangled microwave state, one usually needs the assistance of cryogenic technology to create the low temperature environment. Therefore, the thermal agitating energy can be down to $\sim 10^{-25}$ J when the temperature reaches 8 mK.

The absorption of photons indeed is also important within the atmosphere, however, the attenuation and scattering in the microwave frequency range is much lower than in the optical frequency range. In the case of rain and/or fog, the difference between radar or lidar becomes substantial. For simplicity, we neglect the photon adsorption effect in QRWs since we only wish to study the possible mechanism for the preservation of quantum entangled states instead of the influence of photons absorption. Still, in poor weather conditions, such as rain and/or fog, using lidar or radar can be far less effective than clear weather. This is because the photons may be scattered and absorbed by many free PM2.5 particles in the atmospheric environment.

In this article, we briefly review the development of a quantum radar. We review the ideas based on quantum theory and the prototype of implementing a quantum radar. The QI indeed provided a promising method to make a practical quantum radar. We then use QRWs to simulate the behavior of photons penetrating into the atmosphere and study the possible influence of decoherence interaction from ambient particles. Our QRW method indeed provides a full quantum understanding of the decoherence mechanism between entangled photons and scattering sources. Our result shows that the photon-environment interaction is quite subtle and deserves more careful analysis. For strong decoherence interaction, the entangled photons will soon become decoherent from the collision between photon and environmental particles. However, for the dephasing interaction, the entangled photon pairs will still remain as entangled after collision.

In summary, to properly operate quantum radar, we need to choose the wavelength to be longer enough to avoid absorption and scattering from environmental particles and also shorter enough to reduce the influence of thermal agitation. Moreover, the interaction between ambient particles and entangled photons can only be dephasing-like rather than decoherent-like to sustain the entangled state even after collision occurs. A quantum radar is very challenging to be realized with current technology.

Acknowledgments

The authors thank the Ministry of Science and Technology grant number MOST111-2119-M-033-001. This work is supported by Quantum Information Center in Chung Yuan Christian University and Aerospace Technology Research and Development Center of ROCAF.

References

[1] H. L. Van Trees, *Detection, Estimation, and Modulation Theory, Part III: Radar-Sonar Signal Processing and Gaussian Signals in Noise* (Wiley, New York, 2001).

[2] A. Mallinckrodt and T. Sollenberger, Optimum pulse-time determination, *IRE Trans. Inform. Theory* **3** (1954) 151.

[3] M. I. Skolnik, *Introduction to Radar Systems*, Third Edition (McGraw-Hill, New York, 2002).

[4] Ricardo Gallego Torromé, Nadya Ben Bekhti-Winkel, Peter Knott, *Introduction to Quantum Radar*, https://arxiv.org/abs/2006.14238 (2021).

[5] J. R. Jeffers, N. Imoto and R. Loudon, Quantum optics of traveling-wave attenuators and amplifiers, *Phys. Rev. A* **47** (1993) 3346.

[6] Julia Kempe, *Quantum Random Walks — An Introductory Overview*, arXiv:quant-ph/0303081 (2003).

[7] P. Calabrese and J. L. Cardy, Entanglement entropy and quantum field theory, *J. Stat. Mech.* **0406** (2004), Article P06002, arXiv:hep-th/0405152.

[8] S. Seki and S.-J. Sin, EPR = ER and scattering amplitude as entanglement entropy change, *Phys. Lett. B* **735** (2014) 272.

[9] Robi Peschanski and Shigenori Seki, Entanglement entropy of scattering particles, *Phys. Lett. B* **758** (2016) 89–92.

[10] Michael A. Nielsen and Isaac L. Chuang, *Quantum Computation and Quantum Information*, 10th edition (Cambridge University Press, 2010).

[11] V. Vedral, The role of relative entropy in quantum information theory, *Rev. Mod. Phys.* **74** (2002) 197.

[12] Subhash Kak, Quantum Information and Entropy, *Int. J. Theor. Phys.* **46** (2007) 860–876.

[13] Seth Lloyd, *Science* **321** (2008) 1463; DOI: 10.1126/science.1160627.

[14] E. D. Lopaeva, I. Ruo Berchera, I. P. Degiovanni, S. Olivares, G. Brida and M. Genovese, Experimental realization of quantum illumination, *Phys. Rev. Lett.* **110**(15) (2013) 153603.

[15] S.-H. Tan, B. I. Erkmen, V. Giovannetti, S. Guha, S. Lloyd, L. Maccone, S. Pirandola and J. H. Shapiro, Quantum illumination with Gaussian states, *Phys. Rev. Lett.* **101** (2008) 253601.

[16] S. Barzanjeh, S. Guha, C. Weedbrook, D. Vitali, J. H. Shapiro and S. Pirandalo, Microwave quantum illumination, *Phys. Rev. Lett.* **114** (2015) 080503.

[17] Q. Zhuang, Z. Zhang and J. H. Shapiro, Optimum mixed-state discrimination for noisy entanglement-enhanced sensing, *Phys. Rev. Lett.* **118** (2017) 040801.

[18] Quntao Zhuang, Zheshen Zhang and Jeffrey H. Shapiro, Quantum illumination for enhanced detection of Rayleigh-fading targets, *Phys. Rev. A* **96** (2017) 020302(R).

[19] Ranjith Nair and Mile Gu, Fundamental limits of quantum illumination, *Optica* 7(7) (2020) 771–774.

[20] Quntao Zhuang and Jeffrey H. Shapiro, Ultimate accuracy limit of quantum pulse-compression ranging, *Phys. Rev. Lett.* **128** (2022) 010501.

[21] Jeffrey H. Shapiro, *The Quantum Illumination Story*, https://arxiv.org/abs/1910.12277 (2019).

[22] M. Sanz, K. G. Fedorov, F. Deppe and E. Solano, *Challenges in Open-air Microwave Quantum Communication and Sensing*, https://arxiv.org/abs/1809.02979 (2018).

[23] Marco Lanzagorta, Low-Brightness Quantum Radar, *Proc. SPIE* **9461** (2015) 946113-1.

[24] B. A. E. Saleh and M. C. Teich, *Photonics*, Second Edition (Wiley-Interscience, 2007).

[25] Christopher Gerry, *Introductory Quantum Optics* (Cambridge University Press, 2012).

A Brief Introduction to NISQ Computer and Error Mitigation

生
日
快
乐

L. C. Kwek

Quantum Science and Engineering Centre (QSec),
Nanyang Technological University, 50 Nanyang Ave, 639798, Singapore

Centre for Quantum Technologies, National University of Singapore,
Block S15, 3 Science Drive 2, 117583, Singapore

National Institute of Education, 1 Nanyang Walk, 637616, Singapore

For the last thirty years, we have witnessed a steady progress towards the ultimate realization of a quantum computer. We are not there yet: there are still umpteen challenges and issues with actually realizing a fully programmable computer capable of solving realistic problems with an advantage over classical or traditional computer devices.

Keywords: NISQ; quantum computers; prospect

1. Preamble

In 2004, Tony Woo, then Vice President at the Nanyang Technological University (NTU) and KK decided to set up an Institute for Advanced Studies at NTU. KK raised ten million from the Lee Foundation and the amount was matched with government top-up. Tony and KK then roped me in to help organize the first workshop on spintronics. This was the fairly well-known workshop that Shoucheng Zhang often cited as the platform that led to his fruitful collaboration with the University of Würzburg on quantum spin Hall effect [1].

I subsequently became his Deputy Director, together with Choy Sin Hew, an eminent scholar on the Orchid plant, for the Institute of Advanced Studies (IAS) at NTU. As I have said earlier, the first event that we organized was the spintronics workshop. And at the beginning, there were less than five conferences or workshops a year. KK engaged several people from his successful company, World Scientific Publishing Company, to help with the secretariat. IAS@NTU was a successful venture. With his worldwide connections, KK was able to organize many schools, conferences, workshops, symposiums and public lectures. At its peak, IAS@NTU organized close to twenty conferences and workshops a year. It was a hectic, and sometimes costly period,[a] but for many people whom I know, it was regarded as a golden period for fundamental and applied sciences at NTU. This hectic pace went on for the next thirteen or so years until KK retired. During that period, I was always amazed by KK's energy and ideas.[b]

Fig. 1.   From the left, seating arrangement on the front row at the opening of the School: we have Martial Ducloy, Leticia Cugliondolo, K.K. Phua, His Excellency M. Pierre Buhler, French Ambassador to Singapore, Guaning Su, President, NTU in 2009. The famous blue-hardcover proceedings of 91st session of the Les Houches Physics School in Singapore is shown on the right.

One of the most memorable schools (lasting six weeks) that IAS@NTU organized was the 91st Les Houches "Summer" School in 2009 in Singapore. This was the first, and only, "L'école de physiques des Houches" organized outside the beautiful mountains of the Alps. Although Cécile DeWitt-Morette, the founder of Les Houches School of Physics, could not come for the school, she subsequently came for one of the conferences orga-

---

[a]I had to supplement some of the costs of the events from my own pockets.
[b]He would typically come up with some wonderful ideas at some regular meetings hoping that we would push them through.

nized by IAS@NTU, and I personally made sure that she went home with an IAS@NTU T-shirt. She was a remarkable lady, reticent but very humble and sharp at her age.

Aside from IAS at NTU, KK also roped me into the Tan Kah Kee Foundation and the International Society as well as several school committees. It is through these activities that I saw a different perspective of him: his enthusiasm to support educational projects and the society. He was of course busy with his own company, World Scientific Publishing, but he was also completely immersed into several national projects: the Singapore–China Friendship Association, the National Library Board, and so forth. I must say that silently standing behind KK all this time as a staunch and beautiful supporter was his constant companion and wife, Doreen Liu. Without Doreen's acute business acumen, KK might have squandered his wealth and his business could not have been so successful.

The headquarters of the International Union of Pure and Applied Physics (IUPAP) in 2012 was located under the Institute of Physics (UK) in London. However, it was a period when the post of President was going to be transferred to Bruce McKellar in Australia, and Bruce somewhat felt that Singapore was a better location for him, at least it was closer for him to operate from Singapore. He convinced KK that perhaps IUPAP could move its headquarters to Singapore. KK agreed. So in 2014, Singapore, under IAS@NTU, organized the 28th IUPAP General Assembly in Singapore, and soon thereafter, the IAS@NTU took over the IUPAP secretariat. KK was therefore instrumental in bringing the IUPAP HQ to Singapore and making yet another impact on the research here.

KK's scientific training was in particle physics, specifically, he was interested in the pheonomenological application of Veneziano model to pion-pion scattering [2]. Veneziano proposed a model [3] in the sixties when he observed that the Euler Beta function possesses some nice features: it exhibits the linear Regge trajectories and it shows the essential properties of scattering amplitudes of strongly interacting systems. He also studied forward–backward multiplicity correlations and Chao–Yang Statistics, which were popular and intensively studied in the seventies.

In 2000, KK initiated a conference to celebrate the turn of the mille-nium, and many eminent speakers were invited. In particular, he invited Artur Ekert who subsequently became my boss as the Director of the Center for Quantum Technologies. Artur brought worldwide visibility to this tiny island and made the island a new hub for quantum computation and com-munication. It is this topic that I will turn to next in the remainder of this paper.

2. Quantum Computers

Quantum computers have been a recent (even if I must add that it is already thirty or so years) dream that promises to provide an edge over the computer that we have known for ages, at least for the last seventy or so years based on binary logic. The dream for a computing machine historically can be traced even before the invention of the transistor, in fact as far back as Liebniz in the seventeen century. We are somehow at the early stage of a fully realizable quantum computer. A quantum computer relies on quantum superposition and entanglement to exploit a possible advantage over the current computer. Moreover, it exploits the large Hilbert space to gain greater, and hopefully a more efficient, computational space. The best "programmable" quantum computer to date however boasts of only tens to a hundred qubits (the quantum equivalent of the bits) [4,5] and it is far from competing with any classical computing device. Yet, we know that if we can manipulate a mere 50 "good" (error free) qubits, we could easily beat any current supercomputer. But, we still do not have good control over the errors. A "good" qubit is still a dream. Quantum error correction is almost non-existent experimentally in large enough scale, though in theory, we have plenty of marvelous schemes.

As an intermediate step towards a fully programmable quantum com-puter, John Preskill suggested that we should squeeze our current resources to see if we can achieve quantum advantage or conjure killer applications [6]. In essence, he is asking what we can do in an era where we have only achieved tens to hundreds of noisy qubits, unprotected by quantum error correction. He has coined the word "Noisy Intermediate Scale Quantum" (or NISQ) computers for such devices. There remains the question of how long this period is going to be? By definition, it should not be too long. But, I guess it is hard to tell. Much akin to the time when transistors were invented, most computers, built after December 23, 1947, still rely heavily on fat and clumsy vacuum tubes. Yet, we quickly (around thirty years) began to produce mini-computers for a fraction of the original cost. This started the period of the popularization of computers for hobbyists, professionals and homes. Popularization is needed for a technology to gain acceptance for growth.

Actually even if we need to wait fifty years for the realization of a useful quantum computer, it is a short time span in the history of calculating machines. It takes many years for mankind to move from the abacus to Napier rods to the calculator. Why then should it not take as long? We have been too impatient with our last stride: from renaissance to industrial revolution to modern day (meaning today) "comforts" with wonderful transportation machines to refrigerators to the computer.

In this review, we summarize a modicum of the recent works in this direction. For a more detailed and technical discussion on NISQ computer, we refer the reader to our recent paper [7]. Hopefully, we can also say something about what we can achieve in the near future in this discussion.

3. NISQ Algorithms

The interest in quantum computation started in the 1990s, presumably with Peter Shor. Peter Shor showed in 1995 that one could find the factors of a large integer composed of two large prime numbers [8]. For many people, this observation must have inspired them to look more closely into quantum computers. Not long after, Lou Grover [9] showed that equipped with a quantum computer, one could also search an unstructured database with quadratic speed-up.[c]

Yet, neither the Shor algorithm nor the Grover Search algorithm are regarded as NISQ algorithms. Both algorithms need a fairly sizeable num-

---

[c]I believe that the local community in Singapore began to seriously look into quantum computation in the late 90s. And it started with a journal club with presentations every Friday evening. At that time, NUS worked on Saturdays and a Friday evening was not considered the last night of the week for pubs and bars.

ber of qubits for the algorithms to make any sense. By the end of 2021, even though several magnificient achievements were recorded: the Google Sycamore quantum computer [4], the Jiuzhang all-optical quantum computer [10] and the IBM Eagle quantum computer [5], all of them do not possess sufficient number of qubits needed for a programmable quantum computer. They were all designed to answer the question of quantum advantage, i.e. by analyzing the statistical outcomes of the results from simple algorithms, for example, a circuit with random unitary gates, could one tell that the quantum machines perform better than a classical supercomputer? Moreover, quantum error correction, which is deemed essential for the performance of a fully programmable quantum computer, was nowhere in sight.

The limitation on the number of qubits is not the only problem with current quantum hardwares. Most of the existing quantum computers suffer from some form of interactions with the environment. In short, these quantum computers are quite noisy. Even if these devices show some form of noise resilience, there are many limitations on the type of algorithms that

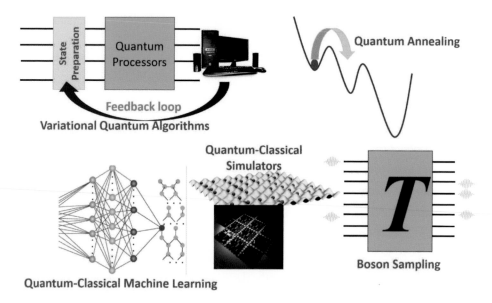

Fig. 2.   Many quantum algorithms, like the Shor algorithm and the Grover search algorithm, are strictly not NISQ algorithms. Yet, it is possible to devise algorithms based on few qubits (up to a few hundreds) that can show quantum advantages. This figure summarizes some of these algorithms.

can be run on such hardwares. Fortunately, there are some algorithms that do not demand a deep circuit nor a large number of qubits. These algorithms are sometimes called noisy intermediate scale quantum (NISQ) algorithms. Some of the well known NISQ algorithms are:

- variational quantum algorithms [11–13],
- quantum annealing [14],
- Gaussian boson sampling [10],
- analog or digital-analog quantum simulators [15–17],
- quantum-classical machine learning [18,19], and
- iterative quantum assisted eigensolver [20–23] and so forth

4. Hardware-Efficient Ansätze

Within NISQ algorithms, there has been extensive study into random parametrized circuits called "hardware efficient ansätze [24] in contrast to the parametrized structured circuits where one has a Hamiltonian at hand, such as unitary coupled cluster or variational quantum eigensolver [11]. This class of hardware-efficient ansatz has been proposed to accommodate device constraints [24]. The common trait of these circuits is the use of a limited set of quantum gates as well as a particular qubit connection topology. In such circuits, trial states are parameterized by quantum gates, tailored to the available physical device [24,25]. The gate set usually consists of a two-qubit entangling gate and up to three single-qubit gates. The circuit is then constructed from blocks of single-qubit gates and entangling gates, which are applied to multiple or all qubits in parallel. Each of these blocks is usually called layer, and the ansatz circuit generally has multiple such layers. The quantum circuit of a hardware-efficient ansatz with L layers is usually given by

$$U(\vec{\theta}) = \prod_{k=1}^{L} U_k(\theta_k) W_k. \tag{1}$$

Here, the variational parameters are given by $\vec{\theta} = (\theta_1, \theta_2, \cdots, \theta_L)$ and, $U_k(\theta_k) = \exp(-i\theta_k V_k)$ is a unitary operator obtained from the Hermitian matrix V_k and W_k describes some non-parametrized quantum gates. The operators V_k are typically single qubit rotation gates, i.e. V_k are Pauli strings acting locally on each qubit. The matrix U_k is a product of single-qubit rotational gates and W_k are unitary entangling gates.

Other types of ansatz have also been proposed [25,26]: alternating layer ansatz, tensor product ansatz [27], checkerboard or brick-layer tensor network [28] and factorized unitary coupled-cluster and adaptive approaches.

5. Error Mitigation

Noise remains the major obstacle towards large scale realization of quantum computers [29,30]. Since quantum error mitigation does not require the encoding of hundreds of qubits as in the normal quantum error correction, it substantially leads to a huge savings of resources, including qubits needed, a feature that is deemed ideal for NISQ simulation. To compensate for computation errors, quantum error mitigation techniques are used at the post-processing stage of the computation [31,32].

Some quantum error mitigation techniques are [33,34]:

- Extrapolation to Zero-error Limit
- Probabilistic Error cancellation
- Virtual Distillation
- Symmetry Verification

5.1. *Extrapolation to zero-error limit*

Extrapolation to zero-error limit is somewhat akin to what Francisco Alcaraz told me many years ago in Tianjin, "first you find what happens for some values of N and then you plot the parameter as a function of $1/N$ and do an extrapolation", except that here, one plots against the estimated error.

Indeed, error mitigation via Richardson extrapolation has been experimentally tested for finding the ground state energy of H_2 and LiH molecules with the variational quantum eigensolver method [35].

Note that a quantum circuit in an open quantum system [36] can be modeled using the Gorini–Kossakowski–Sudarshan–Lindblad (GKSL) equation:

$$\frac{d}{dt}\hat{\rho}(t) = -i\left[\hat{K}(t), \hat{\rho}(t)\right] + \hat{\mathcal{L}}\left[\hat{\rho}(t)\right], \qquad (2)$$

where we set $\hbar = 1$, $\hat{K}(t)$ acts as time-dependent driving Hamiltonian, and $\hat{\mathcal{L}}[.] = \sum_k \Gamma_k(\hat{O}_k[.]\hat{O}_k^\dagger - \frac{1}{2}\{\hat{O}_k\hat{O}_k^\dagger, [.]\})$ is a superoperator. The previous equation describes Markovian dynamics for $\Gamma_k \geq 0$. Whenever the loss rate Γ_k becomes negative [37,38], (2) can also describe non-Markovian dynamics [39,40]. To ensure complete positivity, we impose the constraint $\int_0^t \Gamma(t')dt' > 0, \forall t$. In general, Γ_k are fixed by the nature of the noise affecting the quantum system. For the zero noise extrapolation, we parametrize Γ_k by a dimensionless scalar λ, i.e. $\Gamma_k \rightarrow \lambda\Gamma_k$. For the no-noise case ($\lambda = 0$), the loss term $\hat{\mathcal{L}}[\hat{\rho}(t)]$ in (2) disappears and we get pure unitary dynamics. For $\lambda = 1$, the noise matches the actual quantum device. In summary, zero noise extrapolation involves two steps [7,32,33].

(1) Noise-scaling: we measure several instances of $E[\mu^{(\lambda_j)}]$ for $\lambda_j \geq 1$.
(2) Extrapolation: using the previous measurements, we estimate $E[\mu^{(0)}]$ by extrapolating to $\lambda = 0$.

5.2. *Probabilistic error cancellation*

Given an ensemble of identical states, one can always reconstruct the state measurements on the ensemble. This process is known as quantum state tomography. In quantum tomography [41,42], a quantum state is represented by a density matrix $\hat{\rho}$, and a physical observable or operator is denoted by a Hermitian $\hat{A}$ operator. We use Kraus representation to describe the operation as a map on the state space such that

$$\hat{\mathcal{L}}[\hat{\rho}] = \sum_j \hat{K}_j \hat{\rho} \hat{K}_j^\dagger. \tag{3}$$

Here, $\hat{K}_j$ are Kraus operators. For Markovian dynamics, the GKSL equation in Eq. (2) can always be described with some Kraus operators.

In error cancellation, typically probabilistic, and hence sometimes known as quasiprobability decomposition introduced by Ref. [33], one estimates the expectation value of an observable by sampling from a set of erroneous circuits, labeled $\mathcal{L}_{\text{tot}}^{(l)}$ for $l = 1, 2, \cdots$, such that

$$\langle \hat{A}^{(0)} \rangle = \sum_l q_l \langle A^{(l)} | \mathcal{L}_{\text{tot}}^{(l)} | \rho^{(l)} \rangle. \tag{4}$$

The expectation value of the observable differs from its ideal value due to the presence of noise. Assuming the experimentalist has full knowledge about the specific noise model, the real numbers q_l, which represent quasiprobabilities, can be efficiently obtained. Here, each $\mathcal{L}_{tot}^{(l)}$ represents the total sequence of noisy gates in the lth circuit. It is possible to use Monte Carlo sampling to compute $\langle \hat{A}^{(0)} \rangle$ by randomly choosing the lth circuit with a probability $p_l = |q_l|/C$, where $C = \sum_l |q_l|$. Finally, the final result is computed through the expected value of measurement outcomes $\langle A^{(0)} \rangle = CE[\mu_{eff}]$, where the effective outcome is $\mu_{eff} = \text{sgn}(q_l)\mu^{(l)}$ if the lth circuit is chosen and $\mu^{(l)}$ is the outcome from the lth circuit. The mean value of the outcomes in error cancellation centers around the ideal one with a larger variance due to C.

The error cancellation method described above needs the knowledge of the error model $\mathcal{L}_{tot}^{(l)}$ as is apparent from (4). For practical implementations, it is useful to merge the linearly independent basis set of operations and gate set tomography [43] to remove the impact of localized Markovian errors. This is done typically with a systematic measurement of the effect of the errors to gauge the efficient design of the quantum error mitigation (QEM) circuits. The set of operations including the measurement and single-qubit Clifford gates is shown to be sufficient for the computation of the expectation values of observables. For the single-qubit case, any operation $\mathcal{L}$ is a 4×4 real matrix in the Pauli transfer matrix representation. The latter can be expressed as a linear combination of 16 basic operations $\mathcal{L} = \sum_{i=1}^{16} q_i \mathcal{B}_i^{(0)}$, composed of $\{\pi, H, S, R_x, R_y\}$ gates [43]. Similarly, the same decomposition can be applied to the two-qubit case.

A systematic way to estimate the errors is via gate set tomography (GST). In GST, one tries to mitigate state preparation and measurement errors. The purpose of GST is to determine the noisy individual quantum circuit performance *a priori*. For a single-qubit gate, one begins with the initial states $|0\rangle, |1\rangle, |+_x\rangle$, and $|+_y\rangle$, where $|+_x\rangle$ and $|+_y\rangle$ are the eigenstates of Pauli operators $\hat{\sigma}_x$ and $\hat{\sigma}_y$ with $+1$ eigenvalue, respectively. For noisy devices, these four states are denoted as $\bar{\rho}_1, \bar{\rho}_2, \bar{\rho}_3$ and $\bar{\rho}_4$, accordingly. We also use $\bar{\mathcal{L}}$ (superoperator) to denote a noisy or imperfect gate to be measured. Since we care about the expectation value of physical observables, for the single-qubit case we have observables $\hat{I}, \hat{\sigma}_x, \hat{\sigma}_y, \hat{\sigma}_z$, denoted as $\bar{A}_1, \bar{A}_2, \bar{A}_3, \bar{A}_4$. The mean value of observables, the 4×4 matrix $\tilde{A}$, is nothing but $\tilde{A}_{j,k} = \text{Tr}[\bar{A}_j \tilde{\mathcal{L}} \bar{\rho}_k]$. Similarly, we can also construct the 4×4 matrix g without applying any gate to the initial states as $g_{j,k} = \text{Tr}[\bar{A}_j \bar{\rho}_k]$. This is repeated for each qubit and each single-qubit gate. Statistical estimation of the initial states $\bar{\rho}_k$ and the observables $\bar{A}_j$ are then given by

$$|\hat{\rho}_k\rangle = T_{\bullet,k}, \tag{5}$$

$$\langle \hat{A}_j| = (gT^{-1})_{j,\bullet}, \tag{6}$$

where we note that the hat symbol represents the statistical estimate and $T_{\bullet,k}(T_{j,\bullet})$ denotes the kth column (jth row) of the matrix T, where T is an invertible 4×4 matrix with the following relationship $\hat{\mathcal{L}} = Tg^{-1}\tilde{\mathcal{L}}T^{-1}$. The same procedure applies for the two-qubit case with the only difference being that there is a total of 16 initial states $\bar{\rho}_{k_1} \otimes \bar{\rho}_{k_2}$ and 16 observables $\bar{A}_{j_1} \otimes \bar{A}_{j_2}$ to be measured. Similarly, we have $g = g_1 \otimes g_2$ and $T = T_1 \otimes T_2$. We have to implement a two-qubit gate GST for each qubit pair involved in the quantum program.

5.3. Virtual distillation

Instead of applying error mitigation to a single copy, virtual distillation [44] involves the measurements of M copies of a stat ρ to suppress any incoherent errors. This suppression is done via a measurement of the expectation values $\dfrac{\rho^M}{\text{Tr}\rho^M}$. Virtual distillation estimates the error-free expectation value of an observable O as $\langle O \rangle_{\text{est}} = \dfrac{\text{Tr}(O\rho^M)}{\text{Tr}(\rho^M)}$. For large M, the estimated expectation value converges rapidly to the closest pure state.

5.4. Symmetry verification

Symmetries play crucial roles in physics, and for each symmetry, there exists a conservation law. A key mechanism in symmetry verification involves the measurement of conserved quantities. In symmetry verification [45–47], one is interested to find the ground state or low-lying eigenstates of a Hamiltonian $\hat{H}$ of a system. Suppose the system possesses a symmetry, represented by a unitary operator $\hat{S}$, such that $\left[\hat{H}, \hat{S}\right] = 0$.

Studying the eigenstates of $\hat{H}$ on a quantum computer can be done entirely within a single target eigenspace $\mathcal{S}$ of $\hat{S}$. In a noisy quantum computer, the noise could shift the state outside the target eigenspace, $\mathcal{S}$. By ascertaining if the state of the system remains within the space $\mathcal{S}$ during or at the end of the computation, and discarding all cases where the state leaves the eigenspace, $\mathcal{S}$, one can show that the quantum computation can be made less sensitive to certain types of noise [45]. Symmetry and duality often go together, so, one also wonders if there is a possibility to apply duality for error mitigation.

6. Going Beyond Current Status

In the NISQ era, we anticipate intensive experimental development in quantum hardware design, algorithm design and the search for "killer" applications: more controllable qubits, architecture for deeper circuits and quantum gates with lower but non-zero error rates. There will also be efforts to demonstrate the quantum advantage over classical computers. Like many technological developments, there is no clear-cut distinction when NISQ era will end and herald the Noisy (but) Advanced Intermediate Scale (NAIS) Quantum Computers or the second generation of NISQ computing.

As shown in Fig. 3, we have only just begun our upward climb towards the ultimate goal of fault-tolerant quantum computing. In fault-tolerant quantum computing, quantum information is encoded with multiple physical qubits to generate logical qubits [48–50]. Many popular quantum algorithms, e.g. the Shor factoring algorithm, the Grover search algorithm or even the Harrow–Hassidim–Lloyd algorithm [51] for solving linear equations, rely on error-correcting qubits for their execution. But maybe there are quantum algorithms for these applications that do not really need quantum error correction: Quantum algorithms that can operate with noise.

Living with noise is yet another possibility. It may be possible that we cannot totally eliminate noise, but we can somehow use the noisy environ-

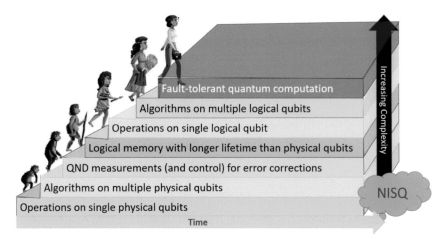

Fig. 3. Figure that illustrates our current progress in quantum computing. Perhaps, we are currently somewhere between the third and fourth steps in the ladder towards a programmable quantum computer. We now have computers that are able to perform error correcting codes for a few qubits. We point out that a progression from a lower to higher stage does not imply that no further work needs to be done to improve the techniques required at the lower stages. For example, improving the control of the operations on single physical qubits should be a goal during the entire timeline of development, even after logical qubits replace physical qubits in computation. Superconducting Qubits are the current leading candidates for the hardware realization of fault-tolerant quantum computation in the long term. Recent demonstrations of high fidelity qubits and gates from major technological players in the industry such as Google and IBM have made superconducting platforms one of the primary choices for demonstrating NISQ era advantage.

ment to our advantage. Modern human beings are well adapted to noisy environment. Our human ear is exposed to a constant basal noise level that we no longer detect nor are we consciously aware. Can logical devices and machines do the same thing?

Overall, we all recognize that there are challenges. It is as if one is asked in 1947 if there is a possibility for a laptop in the "near future". In 1947, the invention of the transistor had just been introduced at the end of the year. Indeed, Thomas Watson, the President of IBM in the forties, did say in 1943 at the dawn of the computer industry that "he [sic] think there is a world market for maybe five computers". We know this is no longer true. There are billions of computers (and supercomputers) around the world now. For that matter, most of our handphones or mobiles are computing machines. So can we have a quantum computer in the next five years? or is it ten years? Or is it twenty years? Or is it fifty years? I am quite confident that we should have a good one in fifty years. The question is: can the period be shorter?

Regardless of the outcomes, one thing I always know for sure: there will be good science coming out of the labs doing quantum computing. And we just need to wait for a spark from one of us to get things running faster.

References

1. Markus Konig, Steffen Wiedmann, Christoph Brune, Andreas Roth, Hartmut Buhmann, Laurens W Molenkamp, Xiao-Liang Qi and Shou-Cheng Zhang. Quantum spin hall insulator state in hgte quantum wells, *Science* **318** (2007) 766–770.
2. H Burkhardt, WN Cottingham and KK Phua, Smoothed veneziano models and applications to pion-pion scattering, *Il Nuovo Cimento A (1965-1970)* **3**(4) (1971) 584–602.
3. G Veneziano, Construction of a crossing-symmetric, regge behaved amplitude for linearly rising trajectories, *Nuovo Cim.* **A57** (1968) 190–197.
4. Frank Arute, Kunal Arya, Ryan Babbush, Dave Bacon, Joseph C Bardin, Rami Barends, Rupak Biswas, Sergio Boixo, Fernando GSL Brandao, David A Buell, *et al.*, Quantum supremacy using a programmable superconducting processor, *Nature* **574**(7779) (2019) 505–510.
5. Jerry Chow, Oliver Dial and Jay Gambetta, IBM quantum breaks the 100-qubit processor barrier, *IBM Research Blog, available in https://research.ibm.com/blog/127-qubit-quantum-process-or-eagle,* 2021.
6. John Preskill, Quantum computing in the NISQ era and beyond, *Quantum* **2** (2018) 79.
7. Kishor Bharti, Alba Cervera-Lierta, Thi Ha Kyaw, Tobias Haug, Sumner Alperin-Lea, Abhinav Anand, Matthias Degroote, Hermanni Heimonen, Jakob S. Kottmann, Tim Menke, Wai-Keong Mok, Sukin Sim, Leong-Chuan Kwek and Alán Aspuru-Guzik, Noisy intermediate-scale quantum algorithms, *Rev. Mod. Phys.* **94** (2022) 015004.
8. Peter W Shor, Polynomial-time algorithms for prime factorization and discrete logarithms on a quantum computer, *SIAM Rev.* **41**(2) (1999) 303–332.
9. Lov K Grover, Quantum mechanics helps in searching for a needle in a haystack, *Phys. Rev. Lett.* **79**(2) (1997) 325.
10. Han-Sen Zhong, Hui Wang, Yu-Hao Deng, Ming-Cheng Chen, Li-Chao Peng, Yi-Han Luo, Jian Qin, Dian Wu, Xing Ding, Yi Hu, *et al.*, Quantum computational advantage using photons, *Science* **370**(6523) (2020) 1460–1463.
11. Alberto Peruzzo, Jarrod McClean, Peter Shadbolt, Man-Hong Yung, Xiao-Qi Zhou, Peter J Love, Alán Aspuru-Guzik and Jeremy L O'brien, A variational eigenvalue solver on a photonic quantum processor, *Nature Commun.* **5**(1) (2014) 1–7.

12. Jarrod R McClean, Jonathan Romero, Ryan Babbush and Alán Aspuru-Guzik, The theory of variational hybrid quantum-classical algorithms, *New J. Phys.* **18**(2) (2016) 023023.

13. Jules Tilly, Hongxiang Chen, Shuxiang Cao, Dario Picozzi, Kanav Setia, Ying Li, Edward Grant, Leonard Wossnig, Ivan Rungger, George H Booth, *et al.*, The variational quantum eigensolver: A review of methods and best practices, arXiv preprint arXiv:2111.05176, 2021.

14. Arnab Das and Bikas K Chakrabarti, Colloquium: Quantum annealing and analog quantum computation, *Rev. Mod. Phys.* **80**(3) (2008) 1061.

15. Javier Argüello-Luengo, Alejandro González-Tudela, Tao Shi, Peter Zoller and J Ignacio Cirac, Analogue quantum chemistry simulation, *Nature* **574**(7777) (2019) 215–218.

16. Philipp Hauke, Fernando M Cucchietti, Luca Tagliacozzo, Ivan Deutsch and Maciej Lewenstein, Can one trust quantum simulators? *Rep. Prog. Phys.* **75**(8) (2012) 082401.

17. Matteo Ippoliti, Kostyantyn Kechedzhi, Roderich Moessner, SL Sondhi and Vedika Khemani, Many-body physics in the NISQ era: Quantum programming a discrete time crystal, *PRX Quantum* **2**(3) (2021) 030346.

18. Giacomo Torlai and Roger G Melko, Machine-learning quantum states in the NISQ era, *Ann. Rev. Condens. Matt. Phys.* **11** (2020) 325–344.

19. Gennaro De Luca, A survey of NISQ era hybrid quantum-classical machine learning research, *Int. J. Artif. Intell. Technol.* **2**(1) (2022) 9–15.

20. Jonathan Wei Zhong Lau, Kishor Bharti, Tobias Haug and Leong Chuan Kwek, Quantum assisted simulation of time dependent Hamiltonians, arXiv preprint arXiv:2101.07677, 2021.

21. Kishor Bharti and Tobias Haug, Iterative quantum-assisted eigensolver, *Phys. Rev. A* **104**(5) (2021) L050401.

22. Zi-Jian Zhang, Thi Ha Kyaw, Jakob S Kottmann, Matthias Degroote and Alán Aspuru-Guzik, Mutual information-assisted adaptive variational quantum eigensolver, *Quantum Sci. Technol.* **6**(3) (2021) 035001.

23. Jonathan Wei Zhong Lau, Tobias Haug, Leong Chuan Kwek and Kishor Bharti, Nisq algorithm for Hamiltonian simulation via truncated Taylor series, *SciPost Phys.* **12**(4) (2022) 122.

24. Abhinav Kandala, Antonio Mezzacapo, Kristan Temme, Maika Takita, Markus Brink, Jerry M Chow and Jay M Gambetta, Hardware-efficient variational quantum eigensolver for small molecules and quantum magnets, *Nature* **549**(7671) (2017) 242–246.

25. Ho Lun Tang, VO Shkolnikov, George S Barron, Harper R Grimsley, Nicholas J Mayhall, Edwin Barnes and Sophia E Economou, qubit-adapt-vqe: An adaptive

algorithm for constructing hardware-efficient ansätze on a quantum processor, *PRX Quantum* **2**(2) (2021) 020310.

26. AV Uvarov and Jacob D Biamonte, On barren plateaus and cost function locality in variational quantum algorithms, *J. Phys. A: Math. Theoret.* **54**(24) (2021) 245301.

27. Kouhei Nakaji and Naoki Yamamoto, Expressibility of the alternating layered ansatz for quantum computation, *Quantum* **5** (2021) 434.

28. Maria Schuld, Ville Bergholm, Christian Gogolin, Josh Izaac and Nathan Killoran, Evaluating analytic gradients on quantum hardware, *Phys. Rev. A* **99**(3) (2019) 032331.

29. Zhenyu Cai, Multi-exponential error extrapolation and combining error mitigation techniques for NISQ applications, *NPJ Quantum Inf.* **7**(1) (2021) 1–12.

30. Suguru Endo, Zhenyu Cai, Simon C Benjamin and Xiao Yuan, Hybrid quantum-classical algorithms and quantum error mitigation, *J. Phys. Soc. Japan* **90**(3) (2021) 032001.

31. Yasunari Suzuki, Suguru Endo, Keisuke Fujii and Yuuki Tokunaga, Quantum error mitigation as a universal error reduction technique: Applications from the NISQ to the fault-tolerant quantum computing eras, *PRX Quantum* **3**(1) (2022) 010345.

32. Ying Li and Simon C Benjamin, Efficient variational quantum simulator incorporating active error minimization, *Phys. Rev. X* **7**(2) (2017) 021050.

33. Kristan Temme, Sergey Bravyi and Jay M Gambetta, Error mitigation for short-depth quantum circuits, *Phys. Rev. Lett.* **119**(18) (2017) 180509.

34. Jarrod R McClean, Mollie E Kimchi-Schwartz, Jonathan Carter and Wibe A De Jong, Hybrid quantum-classical hierarchy for mitigation of decoherence and determination of excited states, *Phys. Rev. A* **95**(4) (2017) 042308.

35. Abhinav Kandala, Kristan Temme, Antonio D Córcoles, Antonio Mezzacapo, Jerry M Chow and Jay M Gambetta, Error mitigation extends the computational reach of a noisy quantum processor, *Nature* **567**(7749) (2019) 491–495.

36. Heinz-Peter Breuer, Francesco Petruccione, *et al.*, *The Theory of Open Quantum Systems* (Oxford University Press on Demand, 2002).

37. CH Fleming and BL Hu, Non-markovian dynamics of open quantum systems: Stochastic equations and their perturbative solutions, *Ann. Phys.* **327**(4) (2012) 1238–1276.

38. Ángel Rivas, Susana F Huelga and Martin B Plenio, Entanglement and non-markovianity of quantum evolutions, *Phys. Rev. Lett.* **105**(5) (2010) 050403.

39. Victor M Bastidas, Thi Ha Kyaw, Jirawat Tangpanitanon, Guillermo Romero, Leong-Chuan Kwek and Dimitris G Angelakis, Floquet stroboscopic divisibility in non-markovian dynamics, *New J. Phys.* **20**(9) (2018) 093004.

40. Thi Ha Kyaw, Victor M Bastidas, Jirawat Tangpanitanon, Guillermo Romero and Leong-Chuan Kwek, Dynamical quantum phase transitions and non-markovian dynamics, *Phys. Rev. A* **101**(1) (2020) 012111.
41. Daniel Greenbaum, Introduction to quantum gate set tomography, arXiv preprint arXiv:1509.02921, 2015.
42. Seth T Merkel, Jay M Gambetta, John A Smolin, Stefano Poletto, Antonio D Córcoles, Blake R Johnson, Colm A Ryan and Matthias Steffen, Self-consistent quantum process tomography, *Phys. Rev. A* **87**(6) (2013) 062119.
43. Suguru Endo, Simon C Benjamin and Ying Li, Practical quantum error mitigation for near-future applications, *Phys. Rev. X* **8**(3) (2018) 031027.
44. William J Huggins, Sam McArdle, Thomas E O'Brien, Joonho Lee, Nicholas C Rubin, Sergio Boixo, K Birgitta Whaley, Ryan Babbush and Jarrod R McClean, Virtual distillation for quantum error mitigation, *Phys. Rev. X* **11**(4) (2021) 041036.
45. Xavi Bonet-Monroig, Ramiro Sagastizabal, M Singh and TE O'Brien, Low-cost error mitigation by symmetry verification, *Phys. Rev. A* **98**(6) (2018) 062339.
46. Chufan Lyu, Xusheng Xu, Manhong Yung and Abolfazl Bayat, Symmetry enhanced variational quantum eigensolver, arXiv preprint arXiv:2203.02444, 2022.
47. Xiaoyu Tang, Chufan Lyu, Junning Li, Xusheng Xu, Man-Hong Yung and Abolfazl Bayat, Variational quantum simulation of long-range interacting systems, arXiv preprint arXiv:2203.14281, 2022.
48. A Robert Calderbank and Peter W Shor, Good quantum error-correcting codes exist, *Phys. Rev. A* **54**(2) (1996) 1098.
49. Emanuel Knill and Raymond Laflamme, Theory of quantum error-correcting codes, *Phys. Rev. A* **55**(2) (1997) 900.
50. Daniel Gottesman, *Stabilizer Codes and Quantum Error Correction*, California Institute of Technology, 1997.
51. Aram W Harrow, Avinatan Hassidim and Seth Lloyd, Quantum algorithm for linear systems of equations, *Phys. Rev. Lett.* **103**(15) (2009) 150502.

Jet Gravitational Waves

生
日
快
乐

Tsvi Piran

Racah Institute for Physics,
The Hebrew University, Jerusalem, 91904, Israel
tsvi.piran@mail.huji.ac.il

The acceleration of a jet to relativistic velocities produces a unique memory type gravitational waves (GW) signal: *Jet-GW*. I discuss here recent results concerning the properties of these GWs and consider their detectability in current and proposed detectors. Classical Jet-GW sources are long and short Gamma-ray bursts as well as hidden jets in core-collapse supernovae. Detection of jet-GWs from these sources will require detectors, such as the proposed BBO, DECIGO and lunar based detectors, that will operate in the deciHz band. The current LVK detectors could detect jet-GWs from a Galactic SGR flare if the flare is sufficiently asymmetric. Once detected these signals would reveal information concerning jet acceleration and collimation that cannot be explored otherwise.

1. Introduction

The detection of gravitational waves from merging compact objects (black holes and neutron stars) binaries opened a new era in Physics and Astronomy [1,2]. The current detectors have been upgraded and new detectors at other wavelengths are being planned. At the same time, new prospects for detection and utilization of gravitational waves from different sources have been proposed. One such less explored source is *gravitational waves from relativistic jets: Jet-GWs.*

Any relativistic jet involves acceleration of mass (or energy density — if the jet is Poynting flux dominated) from rest to a speed approaching the speed of light. The acceleration process produces a memory type gravitational waves signal [3–7] (see Fig. 1). A low frequency detector, will "see"

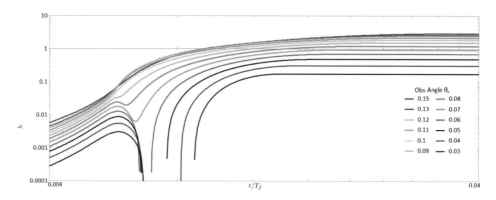

Fig. 1.   The normalized GW ampliture, $h(t)$, for a jet with an opening angle $\theta_j = 0.1$ and a final Lorentz factor, $\Gamma = 100$ for different observer angles. From Ref. [6].

this signal as a sudden jump. A higher frequency detector whose frequency range is comparable to the rise time of this signal, will resolve a temporal structure that can reveal the nature of the jets and the acceleration process. In this article, I review the basic features of *Jet-GWs* and discuss potential sources and detection prospects.

2. Instantaneously Accelerated Point Particle

Consider, first, a point mass m that is instantaneously accelerated to a Lorentz factor Γ. The total energy is $\mathcal{E} = m\Gamma$ (unless specified otherwise, I use units in which $c = G = 1$). While non-physical, this limit gives an excellent idea of the emerging patterns. The waveform is a Heaviside step function:

$$h(t, \theta_v) = h(\theta_v)\mathcal{H}(t),\tag{1}$$

where θ_v is the angle between the direction of the motion of the particle and the line of sight to the distant observer in the observer's frame of reference (see Fig. 2).

The Fourier transform is:

$$\tilde{h}(f, \theta_v) = h(\theta_v)/f.\tag{2}$$

The gravitational wave amplitudes h_+ and $h_\times$ of the two polarization modes are [3]:

$$h^{TT}(\theta_v) = h_+ + ih_\times = \frac{2\mathcal{E}\beta^2}{r} \frac{\sin^2 \theta_v}{1 - \beta \cos \theta_v} e^{2i\phi} , \qquad (3)$$

where ϕ is an azimuthal angle. For a single point-particle, the phase, $2i\phi$, can be ignored. When discussing the metric perturbation of an ensemble of particles, the complex phase leads to a interference and must be taken into account.

The angular dependence of $h(\theta_v)$ is shown in Fig. 3. It exhibits anti-beaming: the GW amplitude vanishes along the jet's direction of motion and it remains small at a cone around it, reaching 50% of the maximal values at an opening angle Γ^{-1}. The function $h(\theta_v)$ attains a maximum of

$$h_{max} = \frac{4G\mathcal{E}}{c^4 r} \frac{\Gamma}{\Gamma + 1}, \quad \text{at } \theta_{max} = \cos^{-1}[\frac{1+\Gamma}{\beta\Gamma}] \approx \sqrt{\frac{2}{\Gamma}}, \qquad (4)$$

where r is the distance to the source and for clarity, Newton's constant, G, and the speed of light, c, have been added to this equation. Interestingly, this relation resembles the classical quadrupole relation for the GW am-

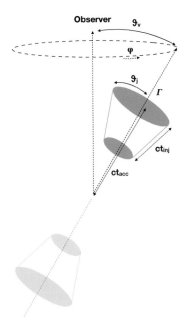

Fig. 2.   A schematic description of the jet. The top shell has reached the final Lorentz factor at a distance ct_{acc} from the origin. The duration of mass injection is t_{inj}. A counter-jet is shown in light colors. From Ref. [7].

plitude $h_{max} \propto (GM/c^2r)(v/c)^2$ (with $v = c$), even though the quadrupole approximation is not valid in this case.

The GW energy is beamed in the forward direction (see Fig. 3). 50% of the GW energy is deposited in a cone with an opening angle $\theta_{50\%} = \sqrt{2/\Gamma}$.

In many cases there are two jets moving in opposite directions. The GW amplitude in this case is the sum of the two signals:

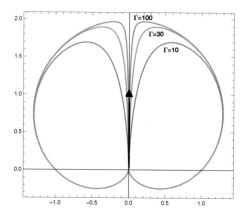

Fig. 3. An antenna pattern for the GW amplitude, h, from an instantaneously accelerated point-like jet for different Lorentz factors. h is anti-beamed with an opening angle Γ^{-1} around the jet direction.

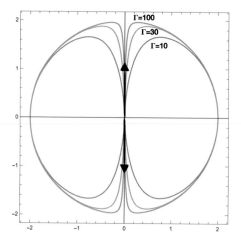

Fig. 4. An antenna pattern for the GW amplitude, h, from two opposite jets for different Lorentz factors. h is almost uniform, apart from being anti-beamed with an opening angle Γ^{-1} around the jets directions.

$$h(\theta_v) = \frac{4\mathcal{E}\beta^2}{r} \frac{1 - \cos^2 \theta_v}{1 - \beta^2 \cos \theta_v^2} , \tag{5}$$

(see Fig. 4). This signal is almost flat apart from two narrow *anti-beams* of width Γ^{-1} along the directions of the jets. The energy is, however, still beamed into two regions of width $\sqrt{2}/\Gamma$ along the jets.[a] Note, that these figures, depicting $h(\theta_v)$ are somewhat misleading. While they depict the correct value of h they ignore a critical factor — the characteristic time scale. This time scale that strongly influences the detectability (as well as the energy flux) may vary with θ_v.

Recalling Eq. (2) we note that formally the total GW energy emitted,

$$E_{GW} = \frac{1}{32\pi} \iint (\frac{\partial h}{\partial t})^2 dt d\Omega = \frac{1}{32\pi} \iint \tilde{h}(f)^2 f^2 df d\Omega , \tag{6}$$

diverges for an instantaneously accelerating particle. This divergence is not physical as it arises from the instantaneous approximation. If the system has a typical timescale t_c then there should be an upper cutoff to the frequency above which $h(f)$ decreases with f much faster than f^{-1} and the integral is finite.

The overall energy emitted is of order [7]:

$$E_{GW} = F(\Gamma)[G\mathcal{E}/c^5 t_c]\mathcal{E} , \tag{7}$$

where $F(\Gamma)$ is a function of the jet's final Lorentz boost. $F(\Gamma)$ vanishes when $\beta \to 0$. When $\Gamma \to \infty$ the situation is more complicated and in extreme cases we have to take into account the details of the acceleration process to properly estimate the energy emitted.

3. Time Scales and the Temporal Structure

The above considerations stress the importance of the characteristic time scale. There are two timescales in the system and the relevant one is, of course, the longest. The two intrinsic time scales are the injection timescale, t_{inj} that is measured in the observer's rest frame (which is the same as the source rest frame). It characterizes the overall duration of the jet. A second

---

[a]This result holds for instantaneous acceleration.

intrinsic time scale, t_{acc}, characterizes the acceleration (it is also measured in the source rest frame).

The signal is determined by t_{acc} via the arrival time from different regions of the jet. The arrival time is related to t_{acc} as:

$$t_o(\theta_v) = (1 - \beta \cos \Delta\theta_v)t_{acc}, \tag{8}$$

where β is the jet's velocity and $\Delta\theta_v = \theta_v + \theta_j$ is the sum of θ_v and the angular width of the jet, θ_j (see Fig. 2). t_o is the time difference from the observation of the beginning of the acceleration and the arrival time of the signal from the furthest point of the jet from observer (see Fig. 2).

The observed time scale of the GW signal, t_c, is the longest between t_{inj} and t_o:

$$t_c(\theta_v) = \max(t_{inj}, t_o) = \max[t_{inj}, (1 - \beta \cos \Delta\theta_v)t_{acc})] . \tag{9}$$

As t_o depends on the viewing angle, the dominant time scale may be t_{inj} for some observers and t_o for others. More specifically, t_{inj} is independent of the viewing angle. $t_o \ll t_{acc}$ for small viewing angles. Thus, unless t_{acc} is very large, $t_c = t_{inj}$ for these angles. However, t_o becomes significantly larger when $\Delta\theta$ increases and for those angles we may have $t_c = t_o$.

The Fourier transform of a memory-type signal that is rising to an asymptotic value $h_0(\theta_v)$ over a timescale t_c, has a general shape:

$$\tilde{h}(f, \theta_v) = \begin{cases} h_0(\theta_v)/f, & f \leq f_c \\ h_0(\theta_v)g(f/f_c)/f_c, & f \geq f_c . \end{cases} \tag{10}$$

The crossover frequency, f_c, in which the Fourier transform of the GW signal changes its behavior is the reciprocal of the time scale, t_c: $f_c \equiv 1/t_c$. The function $g(f/f_c)$ depends on the nature of the source. $g(f/f_c)$ always decreases faster than $(f/f_c)^{-1}$. As the total GW energy must be finite, the integral $\int_0^\infty \dot{h}^2(t)dt = \int_0^\infty f^2\tilde{h}^2(f)df$ yields an asymptotic bound of $g \propto (f/f_c)^{-\alpha_{inf}}$ with $\alpha_{inf} > 3/2$.

The spectral density, $S(f) \equiv \tilde{h}(f) \cdot \sqrt{f}$, is used to characterize the sensitivity of GW detectors.

$$S(f, \theta_v) = \begin{cases} h_0(\theta_v)/f^{1/2}, & f \leq f_c \\ h_0(\theta_v)g(f/f_c)f^{1/2}/f_c, & f \geq f_c . \end{cases} \tag{11}$$

The spectral density at the crossover frequency, $S(f_c) = h_0(\theta_v)/f_c^{1/2}$, is critical to determine the detectability of the signal. The condition

$$S_{det}(f) < S(f_c)(f_c/f)^{1/2} = h_0(\theta_v)/f^{1/2} \quad \text{for some f,} \tag{12}$$

is a necessary condition for detection. For a low-frequency detector (whose typical frequency range is below f_c) this condition is sufficient. However, it is not sufficient for a high frequency detector as $S(f)$ decreases faster than $f^{-1/2}$ above f_c.

A more detailed discussion of the Fourier transform, $\tilde{h}(f, \theta_v)$ and the spectral density is given in Ref. [7]. As an example Fig. 5 depicts $\tilde{h}$ for different jets.

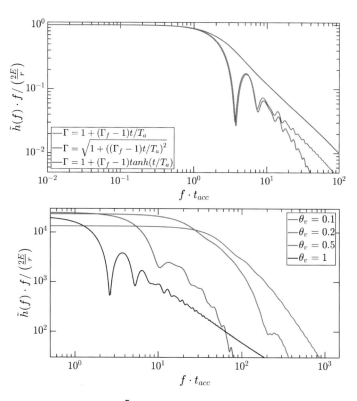

Fig. 5.   (Top) The Fourier transforms, $\tilde{h}$ of the GW signals from jets with three different acceleration models (with $\Gamma_f = 100$, $\theta_j = 0.1$, $\theta_v = 0.9$). In all cases $t_{inj} = 0$. (Bottom) The normalized Fourier transform multiplied by the frequency for jets with the acceleration model $\Gamma \propto R$, based on the numerical code described in Ref. 6 for $\Gamma = 100$ and $\theta_j = 0.1$. Below the crossover frequency, $\tilde{h}(f) \cdot f$ is a constant. From Ref. [7].

4. The Angular Structure

To explore the effect of the angular width of the jet we consider a thin spherical shell ejected simultaneously and accelerated instantaneously. The shell has a half opening angle θ_j and its center is moving at an angle θ_v relative to the observer. If $\theta_j \lesssim \Gamma^{-1}$ the signal converges to the point-particle limit. In the rest, we consider the case of $\Gamma^{-1} \ll \theta_j$, in which cancellations between different parts of the shell become significant and the signal becomes weaker with a broader anti-beaming region around the axis.

We define the observer's line of sight to the emitting source as the z-axis of our coordinate system. The coordinates θ and ϕ are defined in the observer's coordinate system in the usual manner (see Fig. 2). The direction of the center of the jet is $(\theta, \phi) = (\theta_v, 0)$ in the observer's frame of reference. The axial symmetry implies invariance under the transformation $\phi \rightarrow -\phi$. Therefore, the metric perturbation $h_\times$ (which is now summed over the shell) vanishes identically (see Eq. (3), and Fig. 6). In the following, we simply denote $h = h_+$, the only non-vanishing component of the metric perturbation tensor.

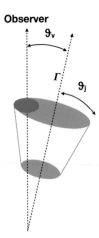

Fig. 6.   A schematic view of the jet (blue) for $\theta_v < \theta_j$. Due to the symmetry, the contribution to the GW amplitude of the part of the jet that is spherically symmetric around the observer (shown in red) vanishes. The amplitude from partial rings with $\theta > \theta_j - \theta_v$, is reduced compared to the amplitude of a point-particle with the same energy and angle to the observer. The jet is symmetric under the transformation $\phi \rightarrow -\phi$: hence, the metric perturbation component $h_\times$ vanishes identically. From Ref. [7].

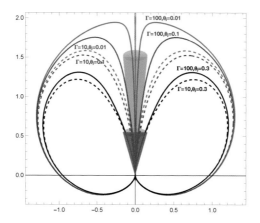

Fig. 7. Antenna pattern for the GW amplitude, h, from an accelerating spherical shell with different opening angles ($\theta_j = 0.01, 0.1, 0.3$ — red, blue and black) and different Lorentz factors ($\Gamma = 10, 100$ — dashed and solid lines). The anti-beaming region is $\approx 0.84\,\theta_j$. The narrow jet with $\theta_j = 0.01 \approx \Gamma^{-1}$ behaves like a point particle. For the larger opening angles ($\theta_j > \Gamma^{-1}$) the pattern is almost independent of Γ.

Integrating over the shell we find:

$$h_{(\theta_v, \theta_j)} = \frac{2\mathcal{E}\beta^2}{r\Delta\Omega} \int_{|\theta_v - \theta_j|}^{min(\theta_j + \theta_v, \pi)} \frac{\sin^3\theta \cdot \sin 2\Delta\phi}{1 - \beta\cos\theta} d\theta \,, \qquad (13)$$

where $\Delta\Omega \equiv 2\pi(1 - \cos\theta_j)$, the solid angle of the cap, and

$$\Delta\phi \equiv \cos^{-1}\left[\frac{\cos\theta_j - \cos\theta_v \cos\theta}{\sin\theta_v \sin\theta}\right]. \qquad (14)$$

Figure 7 depicts the antenna pattern of $h(\theta_v, \theta_j)$ for different opening angles. This angular behavior resembles the point-particle result, with a major difference: the anti-beaming region, which was Γ^{-1} in the point-particle case, is now $\approx 0.84\,\theta_j$ and it is independent of Γ (if $\Gamma^{-1} < \theta_j$). If the jet is wide the signal is also weaker because of the cancellation of the signals from different parts of the shell. For $\theta_v < \theta_j$, only the outer region of the shell, with $\theta > \theta_j - \theta_v$, contributes. The effect is two-fold: regions of the shell with $\theta < \theta_j - \theta_v$ have a vanishing contribution to the amplitude, and even in the outer region, destructive interference between symmetric regions will reduce the GW amplitude. The maximal GW amplitude is now a function of θ_j (compare with Eq. (4)). For small opening angles:

$$h_{max}(\theta_j) \approx \frac{4\mathcal{E}}{r}(1 - \frac{3}{4}\theta_j). \qquad (15)$$

Like in the case of a point source, the GW signal of two-sided jets with angular structure will be the sum of the two signals of the form given by Eq. (13). The amplitude of h will be almost isotropic but now with wider anti-beamed regions along the jets' directions.

5. Detectability

To estimate detectability, we compare the expected spectral density, $S(f)$, to the detector's sensitivity curve that is characterized by $S_{det}(f)$. $S(f)$ is always a decreasing function of the frequency. At low frequencies, namely for $f < f_c$, $S(f) \propto f^{-1/2}$. It decreases faster for $f > f_c$. A typical low-frequency detector will be most sensitive to a jet-GW signal at the lowest end of its frequency response. Namely, the lowest frequency below which S_{det} is steeper than $f^{-1/2}$. At high-frequencies $(f > f_c)$ $S \propto f^{-\alpha}$ with $\alpha > 1/2$. A high-frequency detector is most sensitive for this signal at its lowest frequency below which S_{det} is steeper than $f^{-\alpha}$.

Like almost any GW source involving a relativistic motion, the maximal amplitude of the jet GW is of order

$$h \approx \frac{4G\mathcal{E}}{c^4 r} \approx 10^{-24} \left(\frac{\mathcal{E}}{10^{51} \text{ erg}}\right) \left(\frac{100 \text{Mpc}}{r}\right). \tag{16}$$

$h = 10^{-24}$ is lower than the sensitivity of current GW detectors. Additionally, as we discuss later, these detectors do not operate in the deciHz frequency range that is relevant for most sources. However, we have to remember that the sensitivity of current and planned detectors is much nearer to this signal than the situation in the early 90s when efforts to construct LIGO and Virgo had begun.

For a one-sided jet this estimate is valid for an observer that is at optimal angle, namely at $\theta_v \approx \theta_j$. For two-sided jets, this estimate is valid for most observers apart from those along the jets $(\theta_v < \theta_j)$. Different observers may observe different characteristic crossover frequencies [see Eq. (9)].

6. Potential Sources

Detection horizons can be obtained by comparing the crossover frequency and amplitude of potential sources with the characteristic frequencies and noise amplitudes of several detectors. Given the uncertainties in the signals and in properties of planned detectors this simple approach is sufficient at this stage. Table 1 summarizes the basic features of the different sources and their detection horizon (assuming the high end of the energy range, a viewing angle near the jet and the most suitable frequency) with different GW detectors.

Table 1.   Horizon distances of several current and proposed detectors for different sources. Figure 8 describes the sensitivity curves of LVK, Einstern telscope, BBO and DECIGO. For a moon type detector such as LWGA [10] or LSGA [11] I use sensitivity of $10^{-22}/\sqrt{Hz}$ at 10^{-3} Hz. The values are given using optimistic values of the energy (maximal) and the viewing angle (just outside the anti-beaming cone).

Source	Energy $\mathcal{E}$ (erg)	Frequency f_c (Hz)	Typical Distance	LVK	Einstien Telescope	BBO	DECIGO	Moon [10,11] detectors
LGRB	10^{50-52}	0.1-0.01	6 Gpc	-	-	Gpc	500 Mpc	30 Mpc
sGRB	10^{48-50}	1-10	1 Gpc	-	5 Mpc	75 Mpc	50 Mpc	3 Mpc
Sn Jet	10^{50-51}	0.1-0.01	100 Mpc	-	-	100 Mpc	50 Mpc	3 Mpc
SGR	10^{46-47}	0.001	10 kpc	2.5 kpc	25 kpc	750 kpc	500 kpc	10 kpc

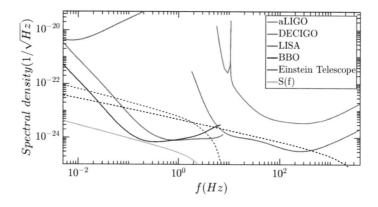

Fig. 8.   The estimated spectral density for a fiducial model [7] of GW170817, $S(f)$, compared with the sensitivity thresholds of GW detectors taken from http://gwplotter.com [8]. The dotted red line shows the GW emission from a CCSN jet with 10^{51} erg at 20 Mpc. The dotted black line corresponds to a Galactic SGR flare of 10^{47} erg at 10kpc. Adapted with modifications from Ref. [7].

6.1. GRBs

GRB jets that have motivated these ideas [3,4] seem to be the most natural sources for these kinds of GWs. These powerful ultra-relativistic jets carry up to 10^{52} erg (for long GRBs) and 10^{50} erg (for short ones). Given the numerous other differences between the two populations (long GRBs are typically detected from larger distances and they have significantly larger t_{inj}) we consider each subgroup independently. The crossover frequency is dominated by t_{inj} for both long and short GRBs. Thus, $f_c = f_l = 0.1–0.01$ Hz for the long and $f_c = f_s = 1–10$ Hz for short GRBs. This frequency range puts the events below the frequency limits of current LIGO–Virgo–Kagra, but around the capability of the planned Big Bang Observer (BBO) [12] and DECIGO [13].

6.1.1. Long GRBs

Long GRBs (LGRBs) are associated with Collapsars — collapsing massive stars. Their typical isotropic equivalent energies range from 10^{51-54} ergs. When taking into account beaming, which is typically of order $\theta_j \approx 0.1 - 0.2$ rad (but highly uncertain) this corresponds to $\mathcal{E} \approx 10^{50-52}$ erg. With typical distances of a few Gpc the GW amplitude from a regular LGRBs is quite low $\sim 10^{-26}$. The duration of LGRBs ranges from 2 sec to a few hundred seconds, with typical duration of $\sim$ 20 sec.

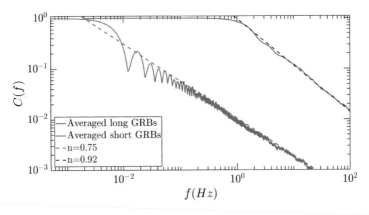

Fig. 9.   The averaged Fourier transform of BATSE's long GRBs [14], vs. that of BATSE's TTE short GRB catalogue [7]. Power law fits for are shown in dashed lines. From Ref. [7].

The average power spectrum of the bursts' γ-rays light curve shows a crossover frequency of $\approx 10^{-2}$ Hz [14]. This crossover frequency reflects the overall duration of the bursts. If the duration of the γ-rays reflects the activity of the central engine (as in the case of internal shocks [9]) this implies that t_{inj} determines the dominant crossover frequency. This frequency is below what can be achieved by any detector built on Earth. It is at the upper end of LISA's[b] bandwidth. However, the expected signal is about thousand times weaker than LISA's planned sensitivity (see Fig. 8). It falls nicely within the band of the proposed BBO and DEFIGO. However, a regular LGRB at a few Gpc is too far to be detected with the sensitivity reach of either BBO or DECIGO.

6.1.2. *Short GRBs and GRB 170817A*

Short GRBs are associated with binary neutron star mergers [15], as demonstrated recently with the observations of GRB 170817A and the associated GW signal GW 170817 [16]. With typical energies of 10^{48-50} erg, sGRBs are much weaker than LGRBs. Correspondingly they are observed from nearer distances, typically of order 1 Gpc, with the nearest ones at distances of a few hundred Mpc (170817A was much nearer than average). Typical sGRB duration is less than 2 sec. The expected crossover frequency (which is again, presumable, is dominated by t_{inj} and not by t_{acc}) is of order 0.5–5 Hz. This is below current terrestrial detectors, it is too low even for the planned frequency band of more advanced detectors like the Einstein telescope[c]. Once more it falls within the planned frequency range of BBO and DECIGO. Again a typical sGRB at a few hundred Mpc is too far to be detected.

The exceptional sGRB, GRB 170817A, was sufficiently nearby that it could have been marginally detected with these proposed deciHz detectors. The properties of this well observed event are well known: $\mathcal{E} \approx 10^{50}$ erg, $\theta_v \approx 20°$, $\theta_j \approx 5°$ [17]. Other parameters, and in particular t_{acc} and t_{inj} that are most relevant for our analysis, are unknown. The injection duration t_{inj}, is not necessarily related to the duration of the observed γ-rays in this event, as those arose from a cocoon shock breakout [18,19]. In the following we assume that $t_{acc} < t_{inj} \approx 1$ sec. Γ, is also unknown but it does not factor into the result as we expect $\Gamma^{-1} < \theta_j \sim 0.1$ rad. Given the viewing angle and the jet angle, GRB 170817A was also ideally positioned in terms of the strength of the GW signal from its jet. That is, we were not within the anti-beamed jet's cone but not too far from it either. The jet-GW of GRB 170717A signal was

---

[b]https://www.elisascience.org/
[c]http://www.et-gw.eu/

not detected. This is not surprising. Figure 8, depicts the spectral density of GRB170817A compared with the sensitivity thresholds of GW detectors [8]. It is below the frequency band of both LIGO and Virgo and also well below their sensitivity. Jet-GW from 170817 would have been detectable by BBO or DECIGO had such detector been operational at the time!

6.2. *Hidden jets in Core-Collapse Supernovae (CCSNe) and low-luminosity GRBs (llGRBs)*

Shortly after the discovery of the first *ll*GRB 980415 (that was associated with SN98bw) it was suggested [20–22] that the emission arose from shock breakout following an energetic jet that was choked deep in the accompanying star. Later on it was realized that, while the detection rate of *ll*GRBs is much lower than that of regular long ones, their actual rate is orders of magnitude larger [23,24]. The detection rate is low because, given their low luminosity, they are detected only from relatively short distances. More recently, it was shown [25] that a significant fraction of CCSNe that are not necessarily associated with GRBs contain energetic ($\sim 10^{51}$ erg) choked relativistic jets. Such a jet deposits all its energy into a cocoon. Upon breakout the cocoon material is observed as a high velocity (0.1–0.2c) material that engulfs the supernova and can be observed within the first few days. Such signatures have been detected [26] as early as 1997 in SN 1997EF and in several other SNe since then. This suggestion was nicely confirmed with the exquisite observations of this high velocity material in SN 2017iuk [27,28]. It is supported by other evidence for jets in SNe [29–33].

If such relativistic jets are associated with a significant fraction of CCSNe then, as the supernova rate is significantly larger than the GRB rate [34], we can expect a significant number of relatively nearby jets that would be sources of jet-GWs. Comparing relativistic CCSNe Jets with GRB jets, we estimate a typical CCSN h to be a factor of 10 larger than the one from a typical long GRB. As we do not have a good clue on t_{inj} we assume that it is of the same order as the one estimated in long GRB, namely a few tens of seconds. Thus, the crossover frequency would be around 0.01 Hz.

If the event is accompanied by a *ll*GRB (arising from the shock breakout of the cocoon) then there will be almost simultaneous EM signal associated with the GW signal (a typical delay would be of order of a dozen seconds). However, if the event is not associated with a *ll*GRB then the EM counterpart in the form of a supernova will emerge a few hours or even days after the GW signal.

6.3. *Giant SGR flares*

Magnetars with magnetic fields larger than 10^{15} Gauss are among the most surprising objects discovered [35]. Magnetars eject giant flares while rearranging their whole magnetic field configuration [36]. Such flares have been observed roughly once per ten years within the Milky Way. The last one, from SGR1806-20 was observed in 2004 [37–40]. It involved the acceleration of $\approx 10^{47}$ erg within a few milliseconds. The corresponding frequency is within the range of current GW detectors. With a Galactic distance the signal might be detectable if the initial injection was beamed. If these flares involve a jet of such energy then they could be detected even with current GW detectors up to distances of a few kpc. Proposed more advanced detectors could observe them from the whole Galaxy or even from M31.

6.4. *Contribution to the GW background*

The relativistic jets that arise from GRBs (both long and short) and hidden jets in SNe produce a background of jet-GW waves at frequency range of ~ 0.01–1 Hz depending on the specific source. Both long and short GRBs are rare and will not make a significant contribution to such a background. However, CCSNe take place at a rate of about one per second in the observable Universe. If a significant fraction of SNe harbor energetic jets the time between two such cosmological events, a few seconds, will be comparable to the characteristic time scale of the GW signals from these jets (assuming that the hidden jets in CCSNe are similar in nature to GRB jets). Depending on the ratio of the time between events and the characteristic frequency of the jet-GW signal we expect either a continuous background, as expected from the GW background from merging binary neutron stars, or a popcorn-like signature, as expected for the GW background from merging binary black holes [41]. With a typical cosmological distance of a few Gpc the corresponding amplitude of this jet-GW background is $h \approx 10^{-26}\mathcal{E}/(10^{51}$ erg).

7. Discussion

The acceleration of a jet to a relativistic velocity generates a unique memory type GW signal. The signal is anti-beamed away from the direction of the jet with an anti-beaming angle of $\max(\Gamma^{-1}, \theta_j)$. Like typical relativistic GW sources, the amplitude is of order $G\mathcal{E}/c^4 r$, corresponding to an amplitude of $h \approx 10^{-24}$ for a jet of 10^{51} erg at 100 Mpc. At low frequencies the signal can be approximated as a step function. This last feature is of course problematic, as it might be difficult to distinguish it from step function type noise sources. At higher frequencies the details of the signal lightcurve could reveal information about the acceleration and collimation of the jet.

Prospects for detection of these jet-GWs depend on the construction of appropriate future detectors. Jets from different kinds of GRBs as well as hidden jets in CCSNe are expected in the deciHz range. Hence most important are detectors like the proposed BBO, DECIGO and several moon detectors (e.g. LWGA, LSGA). Among those sources hidden jets in CCSNe are the best candidates as they are most frequent and as their expected energy is comparable to those of LGRBs.

As jet-GWs are anti-beamed we will most likely miss the γ-rays associated with GRBs from which we will observe these GWs. However, we expect to observe other associated wider signatures such as the afterglow or an accompanying shock breakout signal from the stellar envelope (in case of a LGRB) or from the merger's ejecta (in a sGRB). This shock breakout signal is also most likely to be EM signal associated with a jet-GW arising from hidden jet in CCSN. In this case the actual SN signal may appear hours later.

Table 1 summarizes (under somewhat optimistic assumptions) the detection horizon of different events. Given the large uncertainties I did not attempt to incorporate possible estimates of detection rates as these will be highly uncertain. Detection prospects with current detectors, LIGO, Virgo and Kagra that are operating at frequencies above 10 Hz are slim, unless giant SGR flares involve jets. In that case we could detect a Galactic event even with LVK. Otherwise, we will have to resort to future planned detectors. What is most important is to remember that these signals are memory type. Usually we are not looking for such astrophysical signals. We should make sure to expand GW searches to include such signatures in all future runs. A detection of Jet-GWs would open a new window on jet engines and would reveal features of jet acceleration in the vicinity of black holes that are impossible to explore in any other way.

Acknowledgment

The research was supported by an advanced ERC grant TReX.

References

1. B. P. Abbott, *et al.* and LIGO Scientific Collaboration, Observation of gravitational waves from a binary black hole merger, *Phys. Rev. Lett.* **116**(6) (2016) 061102.
2. B. P. Abbott, *et al.*, LIGO Scientific Collaboration and Virgo Collaboration, GW170817: Observation of gravitational waves from a binary neutron star inspiral, *Phys. Rev. Lett.* **119**(16) (2017) 161101.
3. E. B. Segalis and A. Ori, Emission of gravitational radiation from ultrarelativistic sources, *Phys. Rev. D.* **64**(6) (2001) 064018.
4. T. Piran, Gamma-ray bursts — A primer for relativists, eds. N. T. Bishop and S. D. Maharaj, *General Relativity and Gravitation*, (2002) pp. 259–275.
5. N. Sago, K. Ioka, T. Nakamura and R. Yamazaki, Gravitational wave memory of gamma-ray burst jets, *Phys. Rev. D* **70** (2004) 104012.
6. O. Birnholtz and T. Piran, Gravitational wave memory from gamma ray bursts' jets, *Phys. Rev. D.* **87**(12) (2013) 123007.
7. E. Leiderschneider and T. Piran, Gravitational radiation from accelerating jets, *Phys. Rev. D.* **104**(10) (2021) 104002.
8. C. J. Moore, R. H. Cole and C. P. L. Berry, Gravitational-wave sensitivity curves, *Classical and Quantum Gravity*, **32**(1) (2014) 015014.
9. S. Kobayashi, T. Piran and R. Sari, Can internal shocks produce the variability in gamma-ray bursts?, *ApJ* **90** (1997) 92.
10. J. Harms, F. Ambrosino, L. Angelini, V. Braito, M. Branchesi, E. Brocato, E. Cappellaro, E. Coccia, M. Coughlin, R. Della Ceca, M. Della Valle, C. Dionisio, C. Federico, M. Formisano, A. Frigeri, A. Grado, L. Izzo, A. Marcelli, A. Maselli, M. Olivieri, C. Pernechele, A. Possenti, S. Ronchini, R. Serafinelli, P. Severgnini,

M. Agostini, F. Badaracco, A. Bertolini, L. Betti, M. M. Civitani, C. Collette, S. Covino, S. Dall'Osso, P. D'Avanzo, R. DeSalvo, M. Di Giovanni, M. Focardi, C. Giunchi, J. van Heijningen, N. Khetan, D. Melini, G. Mitri, C. Mow-Lowry, L. Naponiello, V. Noce, G. Oganesyan, E. Pace, H. J. Paik, A. Pajewski, E. Palazzi, M. Pallavicini, G. Pareschi, R. Pozzobon, A. Sharma, G. Spada, R. Stanga, G. Tagliaferri and R. Votta, Lunar gravitational-wave antenna, *ApJ* **910**(1) (2021) 1.

11. S. Katsanevas, P. Bernard, D. Giardini, P. Jousset, P. Mazzali, P. Lognonné, E. Pian, A. Amy, T. Apostolatos, M. Barsuglia, *et al.*, in *Ideas for exploring the Moon with a large European lander* (ESA, 2020), p. 1, URL https://ideas.esa.int/servlet/hype/IMT?documentTableId=45087607031744010&userAction=Browse&templateName=&documentId=a315450fae481074411ef65e4c5b7746

12. J. Crowder and N. J. Cornish, Beyond LISA: Exploring future gravitational wave missions, *Phys. Rev. D.* **72**(8) (2005) 083005.

13. S. Sato et al. The status of DECIGO. *J. Phys. Conf. Series*, **840** (2017) 012010.

14. A. M. Beloborodov, Power density spectra of gamma-ray bursts, *AIP Conf. Proc..* (2000).

15. D. Eichler, M. Livio, T. Piran and D. N. Schramm, Nucleosynthesis, neutrino bursts and gamma-rays from coalescing neutron stars, *Nature* **340** (1989) 126–128.

16. B. P. Abbott, *et al.*, LIGO Scientific Collaboration and Virgo Collaboration, Multi-messenger observations of a binary neutron star merger, *ApJL* **848**(2) (2017) L12.

17. E. Nakar, The electromagnetic counterparts of compact binary mergers, *Phys. Reports* **886** (2020) 1–84.

18. M. M. Kasliwal, E. Nakar, L. P. Singer, D. L. Kaplan, D. O. Cook, A. Van Sistine, R. M. Lau, C. Fremling, O. Gottlieb, J. E. Jencson, S. M. Adams, U. Feindt, K. Hotokezaka, S. Ghosh, D. A. Perley, P. C. Yu, T. Piran, J. R. Allison, G. C. Anupama, A. Balasubramanian, K. W. Bannister, J. Bally, J. Barnes, S. Barway, E. Bellm, V. Bhalerao, D. Bhattacharya, N. Blagorodnova, J. S. Bloom, P. R. Brady, C. Cannella, D. Chatterjee, S. B. Cenko, B. E. Cobb, C. Copperwheat, A. Corsi, K. De, D. Dobie, S. W. K. Emery, P. A. Evans, O. D. Fox, D. A. Frail, C. Frohmaier, A. Goobar, G. Hallinan, F. Harrison, G. Helou, T. Hinderer, A. Y. Q. Ho, A. Horesh, W. H. Ip, R. Itoh, D. Kasen, H. Kim, N. P. M. Kuin, T. Kupfer, C. Lynch, K. Madsen, P. A. Mazzali, A. A. Miller, K. Mooley, T. Murphy, C. C. Ngeow, D. Nichols, S. Nissanke, P. Nugent, E. O. Ofek, H. Qi, R. M. Quimby, S. Rosswog, F. Rusu, E. M. Sadler, P. Schmidt, J. Sollerman, I. Steele, A. R. Williamson, Y. Xu, L. Yan, Y. Yatsu, C. Zhang and W. Zhao, Illuminating gravitational waves: A concordant picture of photons from a neutron star

merger, *Science* **358**(6370) (2017) 1559–1565.

19. O. Gottlieb, E. Nakar, T. Piran and K. Hotokezaka, A cocoon shock breakout as the origin of the γ-ray emission in GW170817, *MNRAS* **479**(1) (2018) 588–600.

20. S. R. Kulkarni, D. A. Frail, M. H. Wieringa, R. D. Ekers, E. M. Sadler, R. M. Wark, J. L. Higdon, E. S. Phinney and J. S. Bloom, Radio emission from the unusual supernova 1998bw and its association with the γ-ray burst of 25 April 1998, *Nature* **395**(6703) (1998) 663–669.

21. A. I. MacFadyen, S. E. Woosley and A. Heger, Supernovae, Jets, and Collapsars, *ApJ* **550**(1) (2001) 410–425.

22. J. C. Tan, C. D. Matzner and C. F. McKee, Trans-relativistic blast waves in supernovae as gamma-ray burst progenitors, *ApJ* **551**(2) (2001) 946–972.

23. A. M. Soderberg, S. R. Kulkarni, E. Nakar, E. Berger, P. B. Cameron, D. B. Fox, D. Frail, A. Gal-Yam, R. Sari, S. B. Cenko, M. Kasliwal, R. A. Chevalier, T. Piran, P. A. Price, B. P. Schmidt, G. Pooley, D. S. Moon, B. E. Penprase, E. Ofek, A. Rau, N. Gehrels, J. A. Nousek, D. N. Burrows, S. E. Persson and P. J. McCarthy, Relativistic ejecta from X-ray flash XRF 060218 and the rate of cosmic explosions, *Nature* **442**(7106) (2006) 1014–1017.

24. O. Bromberg, E. Nakar and T. Piran, Are low-luminosity gamma-ray bursts generated by relativistic jets?, *ApJL* **739**(2) (2011) L55.

25. T. Piran, E. Nakar, P. Mazzali and E. Pian, Relativistic jets in core collapse supernovae, *Astrophys. J.* **871**(2) (2019) L25.

26. P. A. Mazzali, K. Iwamoto and K. Nomoto, A spectroscopic analysis of the energetic type Ic hypernova SN 1997EF, *ApJ* **545**(1) (2000) 407–419.

27. L. Izzo, A. de Ugarte Postigo, K. Maeda, C. C. Thöne, D. A. Kann, M. Della Valle, A. Sagues Carracedo, M. J. Michałowski, P. Schady, S. Schmidl, J. Selsing, R. L. C. Starling, A. Suzuki, K. Bensch, J. Bolmer, S. Campana, Z. Cano, S. Covino, J. P. U. Fynbo, D. H. Hartmann, K. E. Heintz, J. Hjorth, J. Japelj, K. Kamiński, L. Kaper, C. Kouveliotou, M. Krużyński, T. Kwiatkowski, G. Leloudas, A. J. Levan, D. B. Malesani, T. Michałowski, S. Piranomonte, G. Pugliese, A. Rossi, R. Sánchez-Ramírez, S. Schulze, D. Steeghs, N. R. Tanvir, K. Ulaczyk, S. D. Vergani and K. Wiersema, Signatures of a jet cocoon in early spectra of a supernova associated with a γ-ray burst, *Nature* **565**(7739) (2019) 324–327.

28. E. Nakar, Heart of a stellar explosion revealed, *Nature* **565**(7739) (2019) 300–301.

29. T. Totani, A failed gamma-ray burst with dirty energetic jets spirited away? New implications for the gamma-ray burst-supernova connection from SN 2002ap, *ApJ* **598** (2003) 1151–1162.

30. P. A. Mazzali, K. S. Kawabata, K. Maeda, K. Nomoto, A. V. Filippenko, E. Ramirez-Ruiz, S. Benetti, E. Pian, J. Deng, N. Tominaga, Y. Ohyama, M. Iye,

R. J. Foley, T. Matheson, L. Wang and A. Gal-Yam, An asymmetric energetic type Ic supernova viewed off-axis, and a link to gamma ray bursts, *Science* **308** (2005) 1284–1287.

31. K. Maeda, K. Kawabata, P. A. Mazzali, M. Tanaka, S. Valenti, K. Nomoto, T. Hattori, J. Deng, E. Pian, S. Taubenberger, M. Iye, T. Matheson, A. V. Filippenko, K. Aoki, G. Kosugi, Y. Ohyama, T. Sasaki and T. Takata, Asphericity in supernova explosions from late-time spectroscopy, *Science* **319** (2008) 1220.

32. S. Taubenberger, S. Valenti, S. Benetti, E. Cappellaro, M. Della Valle, N. Elias-Rosa, S. Hachinger, W. Hillebrandt, K. Maeda, P. A. Mazzali, A. Pastorello, F. Patat, S. A. Sim and M. Turatto, Nebular emission-line profiles of Type Ib/c supernovae — probing the ejecta asphericity, *MNRAS* **397** (2009) 677–694.

33. A. Grichener and N. Soker, Core collapse supernova remnants with ears, *MNRAS* **468** (2017) 1226–1235.

34. T. Piran, Gamma-ray bursts and the fireball model, *Phys. Rept.* **314** (1999) 575–667.

35. C. Kouveliotou, T. Strohmayer, K. Hurley, J. van Paradijs, M. H. Finger, S. Dieters, P. Woods, C. Thompson and R. C. Duncan, Discovery of a magnetar associated with the soft gamma repeater SGR 1900+14, *ApJL* **510**(2) (1999) L115–L118.

36. D. Eichler, Waiting for the big one: A new class of soft gamma-ray repeater outbursts?, *MNRAS* **335**(4) (2002) 883–886.

37. K. Hurley, S. E. Boggs, D. M. Smith, R. C. Duncan, R. Lin, A. Zoglauer, S. Krucker, G. Hurford, H. Hudson, C. Wigger, W. Hajdas, C. Thompson, I. Mitrofanov, A. Sanin, W. Boynton, C. Fellows, A. von Kienlin, G. Lichti, A. Rau and T. Cline, An exceptionally bright flare from SGR 1806-20 and the origins of short-duration γ-ray bursts, *Nature* **434**(7037) (2005) 1098–1103.

38. S. Mereghetti, D. Götz, A. von Kienlin, A. Rau, G. Lichti, G. Weidenspointner and P. Jean, The first giant flare from SGR 1806-20: Observations using the anticoincidence shield of the spectrometer on INTEGRAL, *ApJL* **624**(2) (2005) L105–L108.

39. D. M. Palmer, S. Barthelmy, N. Gehrels, R. M. Kippen, T. Cayton, C. Kouveliotou, D. Eichler, R. A. M. J. Wijers, P. M. Woods, J. Granot, Y. E. Lyubarsky, E. Ramirez-Ruiz, L. Barbier, M. Chester, J. Cummings, E. E. Fenimore, M. H. Finger, B. M. Gaensler, D. Hullinger, H. Krimm, C. B. Markwardt, J. A. Nousek, A. Parsons, S. Patel, T. Sakamoto, G. Sato, M. Suzuki and J. Tueller, A giant γ-ray flare from the magnetar SGR 1806 - 20, *Nature* **434**(7037) (2005) 1107–1109.

40. T. Terasawa, Y. T. Tanaka, Y. Takei, N. Kawai, A. Yoshida, K. Nomoto, I. Yoshikawa, Y. Saito, Y. Kasaba, T. Takashima, T. Mukai, H. Noda, T. Murakami, K. Watanabe, Y. Muraki, T. Yokoyama and M. Hoshino, Repeated

injections of energy in the first 600ms of the giant flare of SGR1806 - 20, *Nature* **434**(7037) (2005) 1110–1111.

41. B. P. Abbott, *et al.*, LIGO Scientific Collaboration and Virgo Collaboration, GW170817: Implications for the stochastic gravitational-wave background from compact binary coalescences, *Phys. Rev. Lett.* **120**(9) (2018) 091101.

On the Short-Distance Space-Time Structure of String Theory: Personal Recollections

Tamiaki Yoneya[*]

Institute of Physics,
Komaba, University of Tokyo, Japan

Following the invitation to contribute an essay in connection with my paper "On the interpretation of minimal length in string theories" (*Modern Physics Letters A* **4** (1989) 1587), I present a personal overview on string theory, focusing mainly on my own thoughts which led me to the proposal of the 'space-time uncertainty relation'.

1. Earlier Works as Background: An Antinomy

The paper [1] constitutes one of my projects trying to extract possible clues to the ultimate unified theory of quantum gravity from string theory. To make this essay reasonably self-contained, I begin with explaining why I chose such a research direction, sketching the route of my earlier thinkings, despite partially overlapping with a previous memoir [2] published a decade ago.

From my first encounter, as a first-year graduate student, with the then-called 'dual-string model' (DSM, or 'dual-resonance model' DRM) in 1969, a persistent question was how the relationship between the DSM and ordinary quantum field theory (QFT) should be understood: whether the DSM was embodying entirely new physics that went beyond the framework and the principles of usual local quantum field theory (namely, *string theory beyond field theory*, abbreviated as 'SbF') or else the DSM could be derived as a sort of new solution, by means of hitherto undeveloped methodologies, within

[*]Professor Emeritus

the framework of QFT (namely, *string theory within field theory*, abbreviated as 'SwF') These two viewpoints seemed to be completely antinomical to each other. I first envisaged the latter standpoint, taking the picture of the so-called 'fishnet' diagrams as a clue toward their dynamical emergence from a possible field theory for Feynman's (wee) 'partons'. Unfortunately, the problem was too difficult, and I could not attain any useful result.

I then turned to the opposite viewpoint SbF. I thought of two options on this possibility. One of them was the existence of massless spinning states, spin 1 and 2 in open-string (then called the 'Reggeons') and closed-string (the 'Pomerons') sectors, respectively. The existence of massless spin-ning states suggested that the DSM could be regarded as embodying the local gauge principle in a certain generalized sense. In the first commu-nication [3] along this line, I presented an elementary piece of evidence to justify this interpretation: the (on-shell) vertex operators for massless spin 1 states in all ghost-free open string theories, *i.e.* the Veneziano and the Neveu–Schwarz–Ramond models, could be obtained through a mini-mal principle associated with the local gauge transformation of the type $\Psi[x(\sigma),\ldots] \rightarrow e^{i\lambda\left(x(0)\right)}\Psi[x(\sigma),\ldots]$, '$\sigma = 0$' being the end point of an open string, for Nambu's master wave functional [5] ('sting field' in today's ter-minology), in such a way that the (super) Virasoro algebra, and hence the world-sheet (super) conformal symmetry, was satisfied. Thus a would-be extended 'SbF theory' must be regarded as an extension of local gauge the-ories into a new kind of nonlocal field theory which would be characterized by some higher symmetries. In preparing the manuscript of Ref. [3], an idea came to me that the massless spin 2 state of closed string sector should also be interpreted in a similar manner. However, being dominated by a fear that connecting the DSM to gravity must be too weird an idea to pursue, I put aside that possibility: my feverish interests in General Relativity from high-school days were shrinking somewhat after being exposed to the ongoing developments of theoretical particle physics.

The second option was the possibility [4], as proposed by Nakanishi, of crossing-symmetric and Lorentz-invariant decompositions of the gen-eral dual amplitudes, in which each decomposed component exhibits pole singularities corresponding only to a single tree-type Feynman diagram of ordinary field theory; if this decomposition could be reformulated in terms of *second-quantized* string field, it might well be possible to establish an extended non-local covariant field theory. From the early months of 1972 after Ref. [3], I was trying to reformulate the Nakanishi decomposition on the basis of my earlier result [6] about the most general form of Reggeon vertices. I succeeded in extending the method of Ref. [4] that had been restricted to the tree diagrams, to loop contributions [7] at least from the viewpoint of correct counting and smooth matching of the integration regions. But, I could not find a suitable way to conform these results to the concept of the string fields: for some technical reasons in representing the general Reggeon vertex, my rules were encumbered with unphysical extra variables. In retrospect, the trouble was a cumbersome gauge choice [8] in constructing these 'Feynman-like' rules. By this failure, I was forced to retreat from the task of constructing the would-be SbF theory. In 1972 there had been no satisfactory formulation of interacting string dynamics in a clear space-time picture. Mandelstam's light-cone formulation [9] of interacting strings appeared in 1973, and its applications to second-quantization by Kaku-Kikkawa [10] and Cremmer-Gervais [11] appeared in 1974. For the first viable formulation of interacting *covariant* string-field theory, Witten's open string field theory [12], more than a decade after my early attempts was necessary. My attempt was certainly premature.

Furthermore, there had been also a more fundamental difficulty that was inherent in this kind of approaches: it was not possible to separate a different sort of singularities corresponding to the Pomerons, or closed strings, within the framework of my rules for decomposing the pole singularities corresponding to the Reggeons, or open strings. I just mentioned this defect in a footnote of Ref. [7]. Even today, to my knowledge, the problem of a clear characterization of closed-string contributions to exhibit systematically the duality between open and closed strings in covariant open-string field theories has not been appropriately solved.

This experience had been a necessary passage for me to recognize the possible significance of my first anticipation on the connection of closed strings to General Relativity. At first, I thought that formulating concrete evidence for minimal interaction of massless spin 2 state in a similar way as in [3] for open-string models would be the starting point. Then I remembered the work by Neveu and Scherk [13] on the connection of open strings with Yang-Mills theory in the zero-slope limit. I had been made aware of their preprint through a kind letter from Minami, right after completing the manuscript of [3]. So I decided to present an explicit calculation [14] of amplitudes involving gravitons and to confirm its relation to General Relativity in the zero-slope limit, since such a result would be most easily acceptable to all.

In parallel with such calculations, I was considering the possibility of establishing a concrete correspondence principle, by analogy with the formal relationship between commutation relations to Poisson brackets in the quantization of classical mechanics. I could only suggest a modest relation between the generating functionals for tree amplitudes for scatterings of multi-gravitons on both sides [15]. That relation exhibited a remarkable property that the nonlinear effects in General Relativity was replaced by the extendedness of strings. In other words, the existence of higher excitations of a closed string took into account the self-interactions of gravitons: the latter automatically appeared in the zero-slope limit, namely, the limit of local field approximation for strings. The same applied to the case of open strings in connecting to local gauge theory as analyzed systematically in Refs. [15] and [8]. After these works, returning to my first idea of relating closed strings to General Relativity, I had published studies [16] of the coupling of string-graviton to the fermionic strings by extending the method of Ref. [3] to the local Lorentz group and spin connection associated with it.

Meanwhile, I received the paper [17] by Scherk and Schwarz who also discussed the connection of closed strings to General Relativity, from a slightly different viewpoint: in particular, they emphatically stressed that the dual theory should now be interpreted as the fundamental theory of nature not as a theory of hadrons. I was much impressed by their bold stance.

By contrast, my hope towards establishing some basic correspondence principle between string theory and General Relativity had been inclining me toward more formal (and often premature) considerations. I was particularly worried about the antinomy between SwF and SbF: if my results about a derivation of General Relativity from the viewpoint SbF was really meaningful, should I ignore SwF completely for the reason that the ordinary field theory without satisfying Equivalence Principle could not accommodate General Relativity? Alternatively, was there any way for making them reconcile or complement each other? For many years following this period, I had been haunted by this notorious question.

During the second half of the 1970s to the early 80s, I was mainly concentrating on the non-perturbative aspects of gauge field theories, especially the mechanisms of quark confinement, under the influence of the stunning discoveries of various non-perturbative classical solutions and by the proposal of lattice gauge theories. These new developments drove me back to my starting standpoint, SwF. In connection with this, let me mention only one work [18]: I tried to transform the pure $SU(N)$ lattice gauge theory directly into a new kind of string-field theory that could be interpreted as a mean-field theory associated with the Wilson loops. Besides seeking a new approach to the large-N limit, my purpose was to clarify the possible dynamical roles of a non-local global symmetry in terms of the extended string-field like degrees of freedom, and to relate it to the existence of topological string-like excitations. This incidentally provided an early example of the '1-form' symmetry, in today's terminology [19]. The gauge symmetry considered in Ref. [3], on the other hand, had been still a '0-form' symmetry associated with the end points of the open-string fields. Another difference was that this particular version of string-field theory did not seem to respect the world-sheet conformal symmetry which seemed to be crucial to the emergence of massless spin 2 state. I was then encouraged to incline towards SbF again.

2. First Thoughts on the Short Distance Structure in String Theory

The excitement of the 1984 resurgence of perturbative string theory, propelled by the major discovery of the Green–Schwarz mechanism of anomaly cancellation, compelled me further to return to SbF. I now strongly felt that I had to reconsider conceptual problems which had been left unsolved in the 70s. One was the question of geometrical meaning of the string field. I first published small observations [20] seeking for some possible vestige of geometrical structure, or in more concrete terms, background independence, in the framework of the light-cone string-field theory that was the only known acceptable interacting string-field theory at that time: to what extent small deformations of background metric (and later also of the string-coupling constant) could be absorbed into corresponding deformations of the string fields. Meanwhile, a seminal paper [12] by Witten on his covariant open string field theory appeared. In an international workshop on Grand Unification, held at Toyama from where the site of Kamiokande experiment was not far, I gave a review talk on these developments including my own considerations and suggested a simple scenario [21] on how to achieve complete background independence, proposing a purely cubic string-field theory without explicit kinetic terms. This conjecture stimulated a few interesting pursuits [22] along this line by other groups that included some participants in the workshop. More than 35 years have passed since. It seems fair to say, however, that the problem of finding an appropriate conceptual framework for the geometrical interpretation of string fields is still to be solved.

Another problem, remained in the 70s and intimately related to the first one, was nothing but the central theme of the present essay. My anxiety in pursuing SbF was regenerating: to attain a clear correspondence principle that should characterize the significance of the string non-locality contrasting with the local field theory. Such a basic principle, if any, must also account for the resolution of the ultraviolet difficulty of quantum gravity, in addition to bringing classical General Relativity at the long distance regime through producing nonlinearity automatically as discussed in Ref. [15].

Shapiro had already pointed out in the early 70s that the UV finiteness of one-loop closed-string amplitudes was ensured by the symmetry of the amplitudes under the transformation [23] $\tau \to -1/\tau, \tau \to \tau + 1$ of the complex modular parameter τ, the imaginary part of which, up to a proportional constant α', the Regge slope parameter, gives the squared invariant proper length of the trajectory of a closed string forming the loop. This so-called 'modular' invariance allows us to restrict the integration region with respect to τ to the 'fundamental region' $|\tau| > 1$ (and $\mathrm{Im}\,\tau > 0, |\mathrm{Re}\,\tau| < 1/2$). The short-distance region that corresponds to $\mathrm{Im}\,\tau \to 0$ in the case of point particle is then completely eliminated.

This is an extremely superior mechanism for resolving the UV problem since it is based on a symmetry principle rather than on artificial cutoff procedures. However, this type of symmetries is extremely difficult to conform manifestly to the formalism of string-field theories, because the very concept of the string field was intrinsically assuming some particular and restricted manner for foliating the world sheets into strings by choosing some preferential time-like and space-like directions along the sheets, whereas the transformation $\tau \to -1/\tau$ amounts to changing the manner of these foliations discretely (*not* infinitesimaly). I tried hard to forge possible formalisms which would have covariance under the modular transformations in terms of the string field, but couldn't find any positive avenue in that direction. Moreover, essentially the same problem had been there already before going to the loop contributions of string amplitudes, if we would try to formulate the dual transformations for tree amplitudes as symmetry transformations in terms of the string fields. The string fields were not appropriate for this purpose, I thought, in a similar way as the geocentric system had not been appropriate for discovering universal gravitation: probably we needed some fundamental change of our viewpoint.

I thus decided to forget about string-field theory temporarily, and reconsider the problem from a more general and elementary standpoint. I thought that some sort of new uncertainty relation might play a key role: perhaps modular invariance reflected a fundamental restriction on the short-distance structure of space-time itself, in an analogous way as the Planck constant puts a limitation on the smallest volume for classical phase space of coordinates and momenta. Since, in the case of modular transformation, the directions of time and space exchange their roles, it was natural to start from the ordinary time-energy uncertainty relation $\Delta T \Delta E \gtrsim \hbar$. On the other hand, qualitatively speaking, the spatial extension of a string must be of order $\Delta X \sim \alpha' \Delta E / \hbar$ since the string tension as the fundamental and universal physical entity characterizing the dynamics of the string is strictly constant and given by $\hbar/\alpha'$. Then an uncertainty relation directly in the spacetime,

$$\Delta T \Delta X \gtrsim \alpha',\tag{1}$$

immediately followed. That was how I first arrived at this qualitative relation, about the end of 1986. However, I hesitated to publish it: the reason was mainly that I could not provide precise definitions for these uncertainties. In quantum mechanics itself, the status for the time-energy uncertainty relation had been rather problematic, in contrast to the coordinates-momenta uncertainty relation, since the time variable is not a genuine observable.

Despite these or other deficiencies, my conviction was that this relation could be, albeit still at a primitive stage, a crisp but reasonable characterization of string theory. In local field theories, the UV limit, either $\Delta T \to 0$ or $\Delta X \to 0$, necessarily leads to the infinite uncertainty in energies or momenta (or both) and consequently to divergent contributions to any physical amplitudes due to their occupation of infinitely large volumes in the phase space in any physical phenomena. Contrary to this, either ΔX or ΔT increase indefinitely in any would-be UV region, and therefore relevant spacetime regions can never become indefinitely small in string theory. Of course, the difference was the infinite number of degrees of freedom other than the center-of-mass degrees of freedom in string theory: they contribute to expand the spacetime region, *instead of* in the energy-momentum space of the center-of-mass degrees of freedom. This was intuitively just what the modular transformation implies and, hence, the inequality (1) simply extended this property to arbitrary physical processes in string theory, irrespective of the order of perturbation series.

I used to touch on this topic in various seminars and lectures. The reactions were varied: the majority were rather sceptical. Some senior professor who apparently disliked Yukawa's later attempts toward non-local field theories even shouted at me "Do not follow Yukawa's path!" The first instance I had published this idea in a written form was my contribution to a festschrift [24] commemorating the sixtieth birthday of Prof. Kazuhiko Nishijima in 1987, and also had given a brief discussion in a review talk at the second Yukawa memorial symposium later in the same year. Subsequently, I became gradually aware of other independent works concerning the short-distance structure of string theory such as in Refs. [25] and [26] done from different viewpoints.

Then I was beginning to think seriously about the possibility of reinterpreting string theory geometrically such that the theory would appear as a special representation of a theory of quantized spacetime in which time and space could play conjugate roles, analogously at least partially to the canonical coordinate and momentum of phase space prior to quantization introducing $\hbar$. What I had in my mind was an analogy with the Wigner representation of quantum mechanics in which both coordinates and momenta are explicitly used, as opposed to the standard formulation where these two representations are treated mathematically as being mutually exclusive. It had been known that such a non-standard representation of quantum mechanics could be realized using the method of 'deformation quantization' in which the product law in the classical phase space was deformed into the so-called Moyal product. I envisaged, perhaps too ambitiously, that the string theory might be represented as a new and extended sort of Moyal product representation of a quantized spacetime geometry where α' replaces the role of $\hbar$ in the deformation representation of quantum mechanics. I tried very hard for a few months, but, again, I could not find promising and concrete ways to proceed along this line; I had mentioned this possibility as a conjecture in a few internal workshops and invited special lectures at various places around that time.

Now, the work [1] was the first occasion of discussing the result (1) in an academic journal. There I began with emphasizing the general relevance of world-sheet conformal symmetry as the basis for (1): after all, the mathematical essence of the absence of the UV regions in perturbative string theory was that the singularities of Riemann surfaces arise only at the boundaries of their moduli space, and they correspond always to long-distance (thus IR instead of UV) physical processes in the target space-time. During my endeavor toward geometrical formulations, I came to notice that a particular class of the conformal invariant known as the 'extremal length' in the mathematical literature [27] conforms this property and is quite suitable for understanding the relation (1) from the viewpoint of the world-sheet dynamics of strings: although the strict conformal symmetry forbids the notion of length for each separated arc on the Riemann sheet, it is possible to define some analog for lengths for areas which allow to treat a family of permissible arcs lying on them. That also leads naturally to the notion of conjugate pairs of the lengths for arbitrary areas in a certain extremal sense: the conjugate pairs precisely satisfy a reciprocal relation corresponding to (1). Though I could not succeed in the more ambitious project for reinterpreting string theory as the Moyal-type representation of quantized spacetime geometry, it now seemed time to publish my ideas, putting some emphasis on the prospect of spacetime-geometrical formulation as a future possibility. I first sent the manuscript to *Physical Review Letters*, sometime near the end of 1988. To my disappointment, it was rejected. I then submitted it to *Modern Physics Letters A*, the first issue of which I had been receiving as a complimentary copy a few years before.

Over the several years since the publication of (1), I was turning my main research interests to matrix models as toy models for non-perturbative (and non-critical) string theory (or non-perturbative two-dimensional gravity) in lower target spacetime dimensions. My unsuccessful endeavor toward geometrical string theory played, unexpectedly, a useful role here: I was able to apply a similar methodology of the deformation quantization for reformulating [28] a quantum action principle [29] which I had proposed earlier in collaboration with A. Jevicki, setting up a framework of the abstract deformed phase space of the 2D cosmological constant and its conjugate variable, instead of real time and space variables. In lectures at various workshops during this period, I used to emphasize an analogy of the basic non-perturbative equation that took the form of a canonical commutation relation $[P, Q] = 1$ [30], called the 'string equation' at that time, with my uncertainty relation (1) of the critical string theory, although not mentioned explicitly in these papers.

3. Applications to D-branes: A Provisional Unification of SbF and SwF

It seems that the paper [1] had been largely ignored for almost a decade. Only after presenting its application to D-branes, it began to gain some recognition. In the mid 90s, we again experienced a fabulous resurgence of string theory, driven this time by the Polchinski's crucial discovery of D-branes: the D-branes are the carriers of non-vanishing charges with respect to the so-called 'RR'-gauge fields that constitute the supersymmetric multiplets including the gravitational fields. Their physical degrees of freedom are represented explicitly by the end points of open strings: in terms of the world-sheet dynamics of open strings, the spatial directions corresponding to the end point coordinates obeying Dirichlet boundary condition represent movable (transverse) directions, while those obeying Neumann condition represent extended (longitudinal) directions along the trajectory of D-branes including the time direction. Thus the number, p, of spatial directions with Neumann conditions gives the dimensions of the spatial extension of a D-brane that is called a 'Dp-brane'. All the dynamics of D-branes are described by open strings, where the end-point spacetime coordinates work as the collective coordinates, in much the same way as the collective coordinates of soliton-type degrees of freedom in ordinary local field theories. When we consider interactions among D-branes in a low-energy approximation, the spatial transverse distance between two D-branes is characterized by the average length ΔX of open strings, hence longitudinal in the sense of open strings, stretching between them, while ΔT should be identified with the characteristic time-like length scale along the world-volumes of D-brane trajectories. We would then be able to apply the uncertainty relation (1) for the purposes of understanding their dynamics, at least qualitatively, focusing on some important physical length scales involved.

My first work along such a direction was [31], in collaboration with
M. Li, which treated the simplest case of D0-branes or D-'particles',
namely, point-like D-branes with only one (time-like) longitudinal direction, and presented a simple derivation of the characteristic length scales
$\Delta X \sim g_s^{1/3}\sqrt{\alpha'}, \Delta t \sim g_s^{-1/3}\sqrt{\alpha'}$, where $g_s = e^\phi$ is the string coupling constant
related to the background expectation value of dilaton field ϕ, simply by
combining (1) with the ordinary Heisenberg's uncertainty relation for coordinates and momenta. During the preparation of this manuscript, I had
encountered the preprint of the BFSS matrix model [32] that was suggest-
ing an attractive interpretation for the low-energy effective matrix quantum mechanics of D0-branes as a realization in a special light-like frame,
of the so-called M-theory. Subsequently, another proposal [33] of matrix
model (called the IKKT model) associated with D-'instantons' appeared: all
of open-string directions obey Dirichlet condition, and hence a D-instanton
itself was completely localized even in the time direction. In these models, the matrix degrees of freedom represent only the lowest modes (hence,
Yang–Mills and Higgs fields, and their fermionic partners) of open-string
fields stretching among D-particles (BFSS) or D-instantons (IKKT). Being
stimulated by these new types of matrix models, I tried to exhibit close connections of these models with the relation (1): for instance, in the IKKT case
where all directions including time were represented by matrices could be
interpreted as an approximate realization [34] of the commutation relation
corresponding to (1) in terms of matrix coordinate operators, in a way that
manifestly conformed to spacetime Lorentz invariance.

Meanwhile, from the beginning of 1998, an epoch-making work of Maldacena [35] on the AdS/CFT correspondence treating mainly the case of
D3-branes began to propel vigorously the stream of new developments of
holographic correspondences between two different (low-energy) descriptions of D-branes in the framework of supergravity, on one hand, and super
Yang–Mills, on the other hand. The *spacetime* conformal symmetries played
key roles here in establishing a concrete relation, the famous GKP–Witten
relation, between observable quantities on both sides. However, at first
sight, in other cases called the 'dilatonic' cases (including the above matrix

models) where the background dilaton field (and hence the string coupling) was not constant, there seemed to be no such clear symmetry which would play a similar role as in the 'non-dilatonic' cases where the background dilaton is constant (most typically, the case of D3-branes in AdS_5/CFT_4). Through considerations based on the uncertainty relation (1), we pointed out that there were actually extended conformal symmetries [36] even for the dilatonic cases: the main difference was, in the dilatonic cases, the conformal transformations also acted on the dilaton field in addition to other physical fields. For example, the simplest global scale transformation took the form $X_i(t) \rightarrow X_i'(t') = \lambda X_i(t), t \rightarrow t' = \lambda^{-1}t$ and $g_s \rightarrow g_s' = \lambda^3 g_s$, for the spatial transverse coordinates fields X_i in the case of the D0 brane. Notice that the opposite scalings of X_i and t are consistent with (1) and also that the scaling of g_s conforms to the characteristic length scale quoted above. For a comprehensive account of these and related works, see Ref. [37]. Furthermore, the extended conformal symmetry played useful roles in deriving a more detailed correspondence at the level of correlation functions [38] between the 11-dimensional supergravity as the low-energy approximation to M-theory and D0 quantum supermechanics and, much later, its confirmation employing numerical simulations [39].

To conclude, the space-time uncertainty relation seems to be intertwined conceptually with the 'holographic' duality between the two descriptions of D-branes in terms of either gauge-field theory ($\sim$ open strings) or supergravity ($\sim$ closed strings). Recalling my early studies described in Sec. 1, the reader must now remember that the former description is akin to the standpoint SwF, while the latter is related to the SbF, because, the closed-string theory should, after all, be regarded as the UV completion of supergravity as the low-energy effective theory. In that sense, D-branes unexpectedly brought us to a partial reconciliation between SwF and SbF, unifying them as a new duality relation. The uncertainty relation (1), which originated from my anxiety toward a clear correspondence principle from the SbF viewpoint, actually underlies both sides to some extent. The old antinomy that had been annoying me has thus been arriving at a partial and provisional resolution: still 'provisional', since I certainly believe that AdS/CFT or holography (with more recent progress related to information theory) was not the last word yet in our endeavor to explore the meaning of string theory and beyond. To add a broader perspective, I hope, the nature of conjugacy (or non-commutativity) between time and space implied in (1) might be worth deeper contemplation in pursuing the fundamental mysteries of time from an unorthodox but new angle.

Acknowledgments

I am grateful to the editors for providing this opportunity of reflecting upon my spiralling journey of four decades (1970–2010), with Ref. [1] being situated in the middle.

References

[1] T. Yoneya, *Mod. Phys. Lett. A* **4** (1989) 1587.
[2] T. Yoneya, in *The Birth of String Theory*, eds. A. Cappelli *et al.* (Cambridge Univ. Press), p. 459.
[3] T. Yoneya, *Prog. Theor. Phys.* **48** (1972) 616.
[4] N. Nakanishi, *Prog. Theor. Phys.* **45** (1971) 919 and references therein.
[5] Y. Nambu, in *Proceedings of the International Conference on Symmetries and Quark Models*, ed. R. Chand (Gordon and Breach, 1970), p. 269.
[6] T. Yoneya, *Prog. Theor. Phys.* **46** (1971) 1192.
[7] T. Yoneya, *Prog. Theor. Phys.* **48** (1972) 2044.
[8] T. Yoneya, *Prog. Theor. Phys.* **52** (1974) 1355.
[9] S. Mandelstam, *Nucl. Phys. B* **64** (1973) 205.
[10] M. Kaku and K. Kikkawa, *Phys. Rev. D* **10** (1974) 1110; (1974) 1823.
[11] E. Cremmer and J. L. Gervais, *Nucl. Phys. B* **76** (1974) 209.
[12] E. Witten, *Nucl. Phys. B* **268** (1986) 253.
[13] A. Neveu and J. Scherk, *Nucl. Phys. B* **36** (1972) 155.
[14] T. Yoneya, *Lett. Nuovo Cim.* **8** (1973) 951.
[15] T. Yoneya, *Prog. Theor. Phys.* **51** (1974) 1907.
[16] T. Yoneya, *Nuovo Cim. A* **27** (1975) 440; *Prog. Theor. Phys.* **56** (1976) 1310.
[17] J. Scherk and J. H. Schwarz, *Nucl. Phys. B* **81** (1974) 118; *Phys. Lett. B* **57** (1975) 463.
[18] T. Yoneya, *Nucl. Phys. B* **183** (1981) 471.
[19] For a recent review, see, *e.g.* J. McGreevy, arXiv:2204.0304.
[20] T. Yoneya, *Phys. Rev. Lett.* **55** (1985) 1828; *Phys. Lett. B* **197** (1987) 76.

[21] T. Yoneya, in *Seventh Workshop on Grand Unification/ICOBAN'86* (World Scientific, 1987), p. 508.

[22] G. Horowitz, G. T. Lykken, R. Rohm and A. Strominger, *Phys. Rev. Lett.* **57** (1986) 283. See also H. Hata, K. Itoh, T. Kugo, H. Kunitomo and K. Ogawa, *Phys. Lett. B* **175** (1986) 138.

[23] J. A. Shapiro, *Phys. Rev. D* **5** (1972) 1945.

[24] T. Yoneya, in *Wandering in the Fields*, eds. K. Kawarabayashi and A. Ukawa (World Scientific, 1987), p. 419. See also my talk in the 1987 Yukawa Memorial Symposium, in its *Proceedings Quantum String Theory*, eds. N. Kawamoto and T. Kugo (Springer, 1988), p. 23.

[25] D. Gross and P. Mende, *Phys. Lett. B* **197** (1987) 129. See also Gross's talk in *Proc. XXIV Int. Conf. High Energy Physics*, eds. R. Lotthaus and J. Kühn (Springer, 1989).

[26] D. Amati, M. Ciafaloni and G. Veneziano, *Phys. Lett. B* **197** (1987) 81. See also an earlier work, G. Veneziano, *Euro. Phys. Lett.* **2** (1986) 199.

[27] See, *e.g.* L. V. Ahlfors, *Conformal Invariants, Topic in Geometric Function Theory* (McGraw Hill, 1973), Ch. 4.

[28] T. Yoneya, *Comm. Math. Phys.* **144** (1992) 623. See also a related work T. Yoneya, *Int. J. Mod. Phys. A* **7** (1992) 4015.

[29] A. Jevicki and T. Yoneya, *Mod. Phys. Lett. A* **5** (1990) 1615.

[30] M. Douglas, *Phys. Lett. B* **237** (1990) 43.

[31] M. Li and T. Yoneya, *Phys. Rev. Lett.* **78** (1997) 1219.

[32] T. Banks, W. Fischler, S. H. Shenker and L. Susskind, *Phys. Rev. D* **55** (1997) 5112.

[33] N. Ishibashi, H. Kawai, Y. Kitazawa and A. Tsuchiya, *Nucl. Phys. B* **498** (1997) 467.

[34] T. Yoneya, *Prog. Theor. Phys.* **97** (1997) 949.

[35] J. Maldacena, *Adv. Theor. Math. Phys.* **2** (1998) 231.

[36] A. Jevicki and T. Yoneya, *Nucl. Phys. B* **535** (1998) 335.

[37] T. Yoneya, *Prog. Theor. Phys.* **103** (2000) 1081. See also my talk in *Strings MM*, eds. M. J. Duff, J. T. Liu and J. Lu (World Scientific, 2000), p. 305.

[38] Y. Sekino and T. Yoneya, *Nucl. Phys. B* **570** (2000) 174.

[39] M. Hanada, J. Nishimura, Y. Sekino and T. Yoneya, *Phys. Rev. Lett.* **104** (2010) 151601; *JHEP* **12** (2011) 020.

Author Index